Dude, Where's My Ride?

Julian Stan

Gabi Cenco

© 2023 Julian Stan, Gabi Cenco

No part of this publication may be reproduced, stored in a retrieval system, or transmitted in any form or by any means, electronic, mechanical, photocopying, recording, scanning, or otherwise, without the prior written permission of the *Copyright Owner*.

Text: Julian Stan, Gabi Cenco
Editing: Julian Stan, Gabi Cenco
Cover: Julian Stan, Gabi Cenco
Cover Photo: Clem Onojeghuo, Unsplash

ISBN: 9798369692240

Join our Discord: UberDude (not for union members)

https://discord.gg/dCHwjdNKUY

Twitter: @arkdianOfficial & @GabiJustHim
Instagram: @arkdian & @gabicenco
Substack: materiaai.substack.com
www.arkdian.com
www.portocal.app
contact@portocal.app
MATERIA by Julian Stan (podcast)

---

*This book is not medical or financial advice. The authors are not connected in any way to the Uber corporation. The facts in this book are valid as of 2023. The situation may worsen or get better in the future, and some facts might have been missed out due to the complicated state of the private hire industry at the moment.

Bio

Julian Stan and Gabi Cenco are two British/Dacian (Romanian) brothers that moved to the UK in 2013. After a decade long career in the armed forces, they became experts in border tactics, travel documents, and immigration. Their expertise is even more extensive, ranging from regenerative farming, wine making, ranching, and law to immunology/immunotherapies with Bio Products Laboratory. After managing to withstand the intentional destruction of global economies, their businesses are now starting to pick up. The way Uber tackled them financially, has put a temporary halt on their entrepreneurial endeavours in the last three years. Julian is also a serial author, with multiple works published: "This is my Electronica" (a literary ode to Electronic Music and Ibiza, 2010), "Kiss the Girl Save the World Kill the Baddie" (short story non-fiction, 2021), "Coming & Going" (poetry, 2022), "Jazzed Up" (poetry, release in second half of 2023), "Organic Human Health Divine" (health management, release in 2024), and "NFTs In The Wild" (release in 2023, world's first NFT coffee table book containing original digital art). Listen to Julian's podcast, "MATERIA", on Spotify and other streaming platforms. At the time of writing, the brothers have completed over 30,000 trips on Uber and other platforms, and have driven altogether over 400,000+ miles in London and surrounding areas. Julian Stan is qualified in International Affairs, and Gabi Cenco in Law. They both live, work, and do business in London area.

   The two brothers have extensive experience in case building and court representation, having won multiple cases in the UK, of which one in IPO litigation with the Intellectual Property Office in London. The latter case is presented in detail in Julian's 2021 novel.

Other Works From Julian Stan

"This is my Electronica", fiction/non-fiction novel, 2010
"Kiss the Girl Save the World Kill the Baddie", non-fiction novel, 2021
"Coming & Going", poetry, 2022
"Jazzed Up", poetry, 2023
"NFTs In The Wild", coffee table book with NFTs, 2023
"Organic Human Health Divine", health management, 2024

Collaboration:
If you want to place an ad in this book contact us via the channels provided. The cheapest way to purchase this book is via **www.arkdian.com**. Promote *"Dude, where's my ride?"* and earn. Contact us and we'll give you your personalised discount code to give to new buyers. They are saving and you are earning. Once your code is used at www.arkdian.com, we'll pay you straight into your nominated bank account, minus any transfer charges. Tag us on your socials with **#DudeMyRide**.

Published in collaboration with

All that is necessary for evil to triumph is for good men to do nothing.
                    Edmund Burke, 1729-1797

Donations

All donations will aid our legal fight to block any actions that could destabilise the reliable and affordable transportation industry in the future. This concerns riders & drivers alike. Make sure you're using the right address for the currency you're making the donation in.

Bitcoin (BTC)

bc1qgfw9fd3ue05etza4x5dsv3q9
cxusvmqyyxhk0

Ethereum (ETH)

0x3295fb28E00567F453e525458
FEc67b8B4794BCc

Ripple (XRP)

No destination tag required

British Pound (GBP)

International:

IBAN
GB16SRLG60837174991206

BIC
SRLGGB2L

Local Transfer:

Sort Code: 608371
Account Number: 74991206

Contents

1. The Way Uber Used To Be..........**12**
2. How Uber Is Today..........**17**
3. Rumours From Canary Wharf Bankers..........**33**
4. Why It Changed..........**38**
5. Regulators..........**42**
6. The Last Two Big "Secret" Cult Meetings In The UK, Plus A Tercentenary One..........**47**
7. The Green Agenda (Agenda 21 From Rio, etc.)..........**52**
8. The Mayor Of London..........**55**
9. Do Not Harm As Company..........**58**
10. Going Back To The Middle Ages..........**63**
11. Regulators Turning Into Oppressors..........**66**
12. The Supreme Court Involvement.......... **70**
13. The Good Old Taxi Industry..........**73**
14. Uber During The Pandemic..........**77**
15. S.O.S. (The Economy)..........**81**
16. The Silent Voices – Riders..........**86**
17. The Silent Voices –Drivers..........**92**
18. The Silent Voices – MPs..........**97**
19. The Corrupt Unions..........**98**
20. Workers..........**102**
21. Uber's Business Costs..........**106**
22. Uber's Rejected Trips..........**109**
23. The Consequences Of Uber's Imminent Downfall..........**111**
24. Driver Costs Now..........**113**
25. Driver Costs Then..........**116**
26. Driver Sacrifices..........**118**
27. Rider Sacrifices..........**119**
28. Forced Switch To Electric..........**121**
29. Communication With Uber..........**123**
30. What Needed To Change..........**124**
31. What Needed No Change..........**125**
32. The Future For Drivers..........**126**
33. The Future For Riders..........**128**
34. Net Zero Vehicles..........**129**
35. Overregulation Must Stop..........**129**

36. Blocking Anti-Individual, Anti-Freedom Laws for Good……….**130**
37. Private Rides Are Greener Than Public Transport……….**132**
38. What You Can Do As Rider……….**133**
39. What You Can Do As Driver……….**134**
40. Why You Should Do Something……….**135**
41. If You Don't Do Anything……….**136**
42. The Big Audit……….**137**
43. How Uber Planned To Cause Harm To Its Drivers With Clear Intent……….**139**
44. Could Uber Stock Plunge To Under $1 Soon?……….**199**

The Action……….**205**
The Big Conclusion……….**245**
The Legal Way And Who To Contact……….**249**
Last Hour Developments……….**250**

A Campaign To Raise Awareness For PEACE……….**253**

References……….**255**

1. The Way Uber Used To Be

As Uber drivers in London and its neighbouring counties, we were able to enjoy a good work-life balance while earning a comfortable income. This does not include the period between March 2020 – March 2022, for obvious reasons, as there were abnormal market conditions. Back then, we could set our own schedule, choosing when and how often we wanted to work. This allowed us to enjoy weekends off, travel, and invest in other ventures, as well as in our retirement, in more secure ways. We managed to afford to purchase two vehicles, which enabled us to earn even more, as we live in the same house. While driving for Uber, we were able to earn a good wage, accumulating around £1,200 or more in about 45 hours of driving per week (online time). This was significantly more than the minimum wage in the UK, allowing us to thrive financially. We have never earned minimum or less than minimum per economy from the start of our activity on the Uber platform, circa January 2018. We were also able to invest in our health and wellbeing, as well as our mother's, as we had the time and financial resources to do so. The Uber platform provided us with the flexibility to work when it was convenient for us, and we were able to earn more if we got stuck in traffic, or if the journey took longer. All changes agreed with the rider were adding significant money to the journey, in a fair system that benefited both and Uber, even if we were paying 25% service fee for using the Uber platform.

This meant that we were able to earn a higher wage for the time that we put in, which was beneficial for us as drivers. Overall, working as an Uber driver in the London area was a very positive experience for us. We could earn an above average wage, while enjoying a good work-life balance and being able to invest in our health and wellbeing. Although the Supreme Court's decision setting that we are workers was issued in 2021, Uber implemented it from March 2022, disregarding our views completely. As self-employed, which we still are based on laws that apply today and which are above the Supreme Court's decision, we were thriving and having peace of mind, all while affording to enjoy life and save for a new vehicle too. As an Uber driver your vehicle is degrading at an alarming speed, while you don't know when exactly it's going to break down. We weren't worried one bit due to our elevated earnings. Already London's mayor was pushing against private hire drivers to go electric, while helping only black cab drivers through active lobbying. He was lobbying for the ones known for hiding cash from paying tax.

    As Uber drivers in the UK before the company changed their business model, we were able to earn a significant income by working full-time. If we were to drive seven full days, we could easily earn around £6,000 per month each. This was due in part to surge prices, which were higher during times of high demand, or when public transportation was not functioning properly. These surge prices made it possible for us to earn an extraordinary income, and they helped making the choice of driving with Uber a very lucrative venture. The business model of Uber was centred around both drivers and riders. Drivers were earning a good income, and more than 75% of the fares went directly into our pockets. Uber was even offering awesome promotions for completing a certain amount of trips. These bonuses could often be into the hundreds of pounds a week. This meant that we were capable to contribute to the economy by spending most of our earnings on goods and services. Riders were also happy with the service provided by Uber. They could easily book a vehicle quickly using their phones, and they knew that they would be picked up in a timely manner. This made it convenient for them to get around, especially when public transportation was not an option. Stranded women, pregnant women, disabled people, and women with small children had secure transportation to rely on in any situation and at any time, in areas where taxis or public transport were non-existent. Uber was a service for everybody, fixing a dire transportation issue: availability, affordability, and safety.

Uber was so reliable that even MPs, rich international tourists and officials, as well as US senators were using the service while in London. Many of the above individuals were telling me that they had to scrap using black cabs due to poor service and high prices. You, the reader, should not understand that we are trying to bad mouth black cab drivers in London, despite having been spat at, sweared at, and shown obscene signs by many of them during our first years as drivers. There are good and bad individuals on both sides. Some black cab drivers lost their licences in order for that territorial malicious behaviour against another group of hard working individuals to stop. Due to the cashless business model, Uber drivers are more beneficial to the economy than any other cash-based taxi service. I cannot agree with a cashless society for various, some very obvious, reasons but for the sake of understanding the mechanics of the taxi and private hire industries, and their contribution to public finances, this had to be mentioned.

When we first started working on the Uber platform, we were fully employed in a human plasma derived pharmaceutical products laboratory called Bio Products Laboratory. The pay in that lab was so low for the production risks implied, that it made us look for alternatives. After driving for Uber for about two years, we gave up the pharmaceuticals job and started driving full time. Uber was such a good of an opportunity, that people were giving up promising careers in order to earn double or triple the money as a PAYE.

We managed to start our other ventures due to how Uber used to work. Being self-employed, Julian is a writer who managed to invest money in publishing his books, and he also launched an iOS travel app service (Portocal) that is now in danger of collapsing. Uber had improved the lives of millions across the UK, by reducing poverty and diversifying small business, due to the ingenuity of most drivers that chose their own paths, to be their own bosses. This is how all humans should live.

Before the changes, our trip acceptance rate on the Uber platform was never lower than 95%. That means that we were accepting between 95%-100% of the trips offered to us by Uber. When picking up a rider from more than a mile away was something very rare, and riders rarely had to wait for more than five minutes to find an Uber.

Overall, the Uber platform was a win-win for both drivers and riders. Drivers were able to earn a good income and contribute massively to the UK's economy, while riders were getting around quickly and easily. However, the business model of Uber changed in March 2022, after The Supreme Court ruled that all drivers on the platform should be treated as workers rather than independent contractors. This has altered our lives deeply.

Nevertheless, in March 2022, the Supreme Court in the UK ruled that all drivers on the Uber platform should be treated as workers, rather than independent contractors. This decision has likely had an impact on our activity as Uber drivers and the benefits that we were previously enjoying. Before the Supreme Court's ruling, we could work as much or as little as we wanted, and we were able to choose when and where we worked. This gave us a great deal of freedom and flexibility, as we managed to fit our work around our personal life. Having the opportunity to take time off when we needed to, that gave us time to invest in our other ventures, and plan for our retirement.

Back then Uber was even organising round table meetings with select highly rated drivers, in order to discuss what the problems are and how the system could be improved. One of the most important issues Uber should've changed for drivers was to ban aliases on the Uber rider app, and allow only real and complete names. The round table meetings were all about fixing the service and making it better. It was one of the best approaches in the private hire industry.

2. How Uber Is Today

Those good times are now gone, as a general conspiracy against people thriving, living a good life, and moving around as we please is unfolding in every aspect of our lives. Corporations that have been on the outside of the agenda, have now been taken over and steered away from improving life, while governments, instead of rolling back their intervention into private life, they want full control of everybody and everything. It's not the government but the corporations, and vice-versa. They all work together against me and you.

In March 2022, Uber implemented the Supreme Court's decision to treat its drivers as workers and not as independent contractors. An Uber driver can never be a worker based on existing laws, and Uber could've chosen not to apply the ruling, as it is less than a law. It has the statute of law, but by not having been through Parliament, existing laws prevail. Uber was free to keep the old model. They chose to change as they saw a massive opportunity to make more money, while its current management have a history in destroying business models and not caring about anyone but themselves and the agenda they work for. If you think this is a conspiracy theory, you need to watch interviews with the founder of the World Economic Forum, during which he's explaining how they took control of corporations and governments worldwide. Don't be uninformed! You have a responsibility to know what is going on around you and with your life. This is what you need to become in life, an all-knowing individual. This is the only way to end conflicts between each other.

We'll explain briefly what we meant above. The current manager of Uber for London used to work for British Airways. If you've flown with them, you should know that before roughly 2018 you had at least a drink and snacks included in the flight price. A rupture happened in this practice and now, although you're paying way more to fly them compared to low cost airlines, you're no longer getting anything, and you have to pay for everything onboard. You're not even allowed to have your own alcoholic drinks if you want to. This is despicable, and most probably Uber's current manager had something to do with that. You always look at patterns in the businesses run by certain individuals. And this manager had quite some jobs in the past… As an Uber driver, as mentioned previously, you have all sorts of people as riders, and many are very well positioned in society. We even had union representatives from British Airways, and the way they're describing BA's leaders is shocking. Whenever they negotiate with each other, BA's leaders are "heartless and care about nobody but themselves, while having the power and plenty of resources to fix everything". They used words like "vicious" and "cruel" when describing them. When intelligent and important people are willing to share these details with a stranger driver, it does make you think a little and pushes you to do more research. When you also see the same behaviour in your line of work, it's quite clear, isn't it?

At peak times (February-December) each of us is rejecting more than 3,000 trip requests in a 30 day period. According to Uber's rough estimates of about 45,000 drivers being active on their platform, there are roughly 150,000,000 (one hundred and fifty million) trip requests being rejected by Uber drivers in London area alone per month, at peak time. We said 150 million because we've seen drivers that reject more than us. This is a staggering number that you are not allowed to know. Some riders have to wait even for more than two hours to get a vehicle, if the driver doesn't decide to cancel on the way, after realising that he is earning less than £0.50/mile for that trip. Uber used to pay at least £1.25 per mile before the changes, and while now that price is officially £1.09 (a 12.8% decrease per mile), it can even drop to less than £0.40 per mile in some instances. There is no consideration for the distance we have to drive to pickup location, and in many instances Uber wants us to drive 10-15 miles overall for just £4.30. It's as if you can teleport yourself and the car to the pickup location. No matter what, today Uber wants drivers to drive a lot of miles, whether it's during the trip or to the pickup location or both.

If that isn't enough, surge prices today are lower than non-surge prices in 2019. This has decreased drivers' overall income with 50%-60%. Drivers have to work almost seven days a week, compared to optional weekends before the changes. Even before, Uber was trying to manage a driver's weekly income, but now they are limiting it so much that it is almost impossible to earn £1,000 per week, even if you're online for 60-70 hours, and all while the demand is at least 20%-30% higher than before the pandemic. The demand is not due to a lack of Uber drivers! There are plenty of drivers all around you, they're just rejecting the trip requests as they are not financially viable for them.

If everything you know about what is happening with Uber comes from official channels or the media, everything you know is wrong.

Transportation should always be good money. Riders are now paying more than before and drivers earn less than before. This decision was a perfect opportunity for Uber, and a disaster for drivers and riders, while the latter two blame each other. In order to obey the court's decision Uber can now keep for themselves somewhere between 30%-52% from what the rider is paying. The driver is being kept in the dark about what riders are paying, compared with before. Uber is saying that they need to charge more to fulfil the decisions's requirements, to pay at least minimum per hour (trip time), holiday pay, pension, and for pension manager (Adecco). I do not want to know how much is Uber paying (we are actually paying) for that contract with Adecco. We have rejected all these benefits but Uber is still charging our riders the percentage to pay other drivers. The only benefit we could not reject was the holiday pay, which is, as per the T&C, 12.07% of the previous week's earnings minus running costs (which makes no sense as the driver is paying for those). No driver knows how Uber is doing these calculations and from what, as every time you try to understand by doing the calculations yourself, you get different numbers than what Uber is paying you. It is a painstakingly complicated algorithm that need not be this way. The way it used to be worked seamlessly.

It is an enigma why the Supreme Court of the UK decided that we are workers, when the law states that we can never be workers. On top of that, hackney carriage taxi drivers working with platforms identical to Uber are not workers! There has been a case with a taxi driver asking for benefits, and the same Employment Tribunal of the Supreme Court decided that he is not a worker. It is mind boggling and the courts have brought about such an entanglement in the interpretation of simple laws, that it contradicts simple logic and disturbs the entire legal environment! This has external influence written all over it, as again, you look at the patterns in society. The Supreme Court's decision matches the general global narrative of making it harder and harder to earn a living for the individual. All for the sake of benefits that no one needed. There is a black hole worth trillions of dollars in global pension funds that people don't know about. They will eventually crash!

We have never earned minimum or less than minimum per economy on the Uber platform. We did not need holiday pay as we were earning over ten times more than the weekly 12.07% from trip fares. We do not believe in pension, as that is money we will never see. If you know what is going on underneath this fake image of a safe financial environment, how it's being damaged by the government, you now that all pension funds are in grave danger and that they will eventually crash. We cannot trust any institution, hence we always look to manage our finances ourselves. We were earning over 2x an average wage, around £5,000 per month, without working seven days a week like nowadays, and this was enough for us to manage our own private portfolios. We did invest a lot during those years, and most of it in our own physical health.

Nowadays Uber is lying to drivers, just to make them bite and accept the trips. They might show the driver that the specific trip is being sent to them with a surge price (nothing compared to before) and at the end of the trip the driver receives a non-surge price, and lower than what was shown on the trip card offer when the driver accepted the trip. There is no way to get that back, as Uber is training its customer support agents to solve nothing. They have a pitch that they are presenting you with as if they are bots that don't even understand English. Every time we interact with them they leave us confused and baffled.

Of course Uber is complying with this malicious push for electrification, without any financial support for drivers and severely decreased income. They are even lying about the running costs of an electric vehicle, promoting that it is cheaper than a hybrid, which is an abomination. To run an electric vehicle every day, for less than 10 hours of driving on the platform per day, it will cost about £1,200 to charge per month (2022 real world data/ having driven one). A hybrid is costing about £600 per month for the same driving time per day, without altering your life around charging the vehicle. They are not even green vehicles, but the opposite, having a production and running toxicity of more than 10x compared to a full diesel, which could become an almost zero emission vehicle with a simple hydrogen on demand device mounted on the engine. Feel free to research this topic further.

If you didn't understand how our income dropped by 50%-60% compared to before while the demand has increased, nowadays we have to work almost seven days a week and struggle to make £800-£900 per week, inclusive of the holiday benefit of 12.07% of the previous week. This is a rough estimation but it can be clearly seen that actually the decrease in income can actually be higher.

Considering the running costs of a taxi driver, this is a straight on tragedy! The authorities and Uber know all about this but they want to keep everything under the carpet. From a company fighting this type of dangerous overregulation, Uber turned into a supporter of Transport for London and of all these policies that will damage human health and the environment exponentially more. Add to this the intentional crashing of the economy, and you have millions of people negatively impacted by only one category of drivers.

Then there are the changes in central London, which make it impossible for drivers (no matter the vehicle driven) to make a living. Full petrol/diesel and Hybrid owners are paying the congestion charge (£15 per day, five days a week between 7am-6pm, 12pm-6pm on weekends). On top of that, with all the speed restrictions and cycle lanes, speed tracking cameras, speed bumps and humps, short trips can take ages. Even when driving a mile for £4.30 it takes one hour, Uber might pay you at the end £4.69, as example, or the same, £4.30. So, in central London you cannot make money anymore, based on the current algorithm on the Uber platform. When we contact Uber they're always saying that the trip did not match the distance and time thresholds to be paid more. This wasn't a problem in the past, as independent contractor.

Uber is a Big Data company, therefore they are recording every second of any trip from all the angles.

Today, Uber is saying that the contract is no longer between us and our riders, but between Uber and the rider, therefore Uber is no longer required to disclose to the driver how much is the rider paying for the trip. This has opened the door to massive fraud compared to before. Even in 2019 there were instances when we knew that Uber kept more than 25% from the fares, and in some instances even from our tips. From partial transparency Uber is now offering zero transparency to drivers. The mayor of London has supported these specific changes, as he knew what the impact on the industry will be. This situation has been ongoing for a year now, and the authorities, the regulators, the unions (that have actually created this mess) are completely deaf to the cries of private hire drivers driving on ride hailing platform like Uber, etc.

Extreme misinformation about the wellbeing of Uber drivers under the new rules has created an unprecedented dangerous environment for human survival, as drivers have no voice.

As drivers, we can no longer work when we want, because we have to work almost every day, we can no longer work as much as we want because we have to work a lot to survive (60-70 hours per week), we can no longer earn as much as we want because there is no opportunity anymore (no matter if we are the only drivers in an area with huge demand), and we can no longer work where we want because Uber is not increasing the pay if you're stuck in traffic (or in case of significant detour due to road closures).

The opportunity was that if you were the only driver on a five mile radius and ten riders wanted to book an Uber, the one willing to pay the most was booking you. It was an auction, free market. Nowadays you are the only driver in the area and 100 people want to book you, and Uber is still sending you the trip request with £4.30 until you accept it, or leave the area really pissed off. This is how you know that Uber is no longer caring about riders either, as it prefers they don't get cars, than pay drivers more. Uber might even pump the price more, between driver rejections, and the driver doesn't know a thing. Uber, the standard in ride hailing platforms, is now a complete mess, intentionally.

There are even instances when Uber is sending a trip to a driver worth £11.15, as shown on the trip card offer, and after the trip is completed the wallet shows a total of £11.14. So, £0.01 out of millions of trips equals to millions of pounds in Uber's pockets, from fraud. There is no way a driver can challenge Uber for a full audit. It is money lost and a lot of headaches for the driver. It is not a great loss for the driver, but it is a matter of fairness, transparency and morality. It is not something the drivers agreed on.

All this extra money withheld by Uber starting March 2022 is not coming back to the driver under the shape of the so called benefits. It is lost income! Uber is just keeping and managing it for the driver, when that is money that should've been in the driver's pocket. So, what is a benefit if you're bleeding 3x-4x more money overall, maybe even more?

We're not saying Uber cannot sometimes keep less than 30% of the total fare but that happens in extreme situations, when in fact you should receive an increased surge price. When a lot of people are trying to book an Uber and you're the only driver on a five mile radius, yes, as an available driver you should be able to earn £50 per hour, because at least one of the riders has more important business they need to attend to compared to all others. For some it can mean being late to business meetings, which could translate into losing contracts worth millions of pounds.

What about being paid £4.30 for a two mile ride when you have to pick up four massive lads, maybe weighing over half a tonne in total, and your impeccable new vehicle starts creaking as they struggle to fit in? Uber would argue that the one making the booking is allowed to invite friends. Is this fair for a driver? Uber is an affordable service but it should be fair too, for all parties.

There are no night fares, nor extreme weather fares anymore. Before, the surge was compensating for everything. Now, at 2:00 AM, £4.30 for 10 miles overall, snow and icy, £4.30 for 15 miles overall, heavy fog, £4.19 for seven miles overall, heavy rain & hail, £4.30 for nine miles overall, etc. If you're a driver you may know how difficult these driving conditions can be, and working late hours should entail very high rewards.

Uber now says that delays and traffic conditions are now included in the shown price, based on their data, which can never be accurate. Their systems can never know what is actually going to happen in the future, after starting the trip. But because they are withholding in excess of 50% or more from the fares, thus many times being expensive for the rider, there is no leverage to pay the driver more once they've accepted the trip request. Uber does not want to lose a penny from their share for the benefit of the driver, who's actually facing an uncertain outcome of the ongoing trip and traffic predicament.

There is a model that used to work, and that is the previous model. That is a precedent for what Uber is now calling impossible to manage in another way. To please overregulation, everyone needs to suffer and earn less. It's the beautiful future of being governed, you see.

Uber still says the driver will see a reduction of 25% from the total fare, as before, although they pay 1.09 per mile or less, which is an unbelievably irresponsible argument to promote, in the context of them withholding way more than 30% from the total fares most times. Before, they were paying a minimum of £1.25 per mile. It's like they don't know how to do math. They don't care anymore.

They're now referring to the same thing in a different way. **How it used to be**: Uber was charging the rider, keeping 25%, and paying the driver the rest of the fare, plus tips, promotions, fees, etc. **How it is**

today: Uber is charging the rider, keeping as much as they want (30%-52%), and paying the driver the rest of the fare, plus tips, promotions (almost non-existent or severely conditioned), fees, etc. There is intent in what Uber is doing today, and that's the aggravating factor.

They are now being charged VAT, and that is absurd. It means that Uber should purchase the service from its drivers on drivers' terms. Otherwise, what is the value added to the service? And drivers are paying the VAT, not Uber, as they keep substantially more from the riders' fares. We spoke with VAT lawyers in London and they confirmed that it is outrageous that the government is charging VAT to companies like Uber. It is wrong and meant only to destabilise drivers and the service!

Drivers can cash out daily. When we cash out Uber takes 0.50 and put your account on -0.50. Next day you start with -0.50, and when you cash out again they take another 0.50, so in total they charge a total of £1 for one cash out. This happens more often than not, so you lose an extra 0.50. There is no way to challenge Uber as they'll accuse you of fraud.

Uber is banned from areas around London by certain councils, and the locals are complaining about not having transport into London, just to protect local businesses. Uber drivers are local business, and today more money is being lost from the economy through various means as never before.

Proof that there is a certain pressure on Uber to stop people from going places by using the platform is the fact that no matter how many drivers reject one trip, they will not pump the price up, even if the rider is willing to pay more just to get there.

We'd love to run Uber ourselves for Europe and the UK. It would become the most amazing service ever. Unfortunately that is not possible.

Due to Uber's practices a lot of money is being left on the table. Less riders can book an Uber in the same amount of time, although all riders want to book and should be able to book. There are always plenty of drivers around you, unless you're booking from a remote area.

Each company is a mirror of their leadership. There lies the problem. We call it the three "C": Corruption, Collusion, Carelessness. You might say corruption is a characteristic of public institutions, but you're wrong. A big company can be corrupted by something, an idea, a group, a person, a creed. It just doesn't fall under the criminal law until it is proven that it causes harm in society.

Uber is also setting a £20 cap for tips. If you, the rider want to give more tip to the driver, because you liked him/her, maybe you're a millionaire and want to offer that driver the opportunity to not need to drive for a living anymore, you want to change that driver's life, you cannot tip as much as you want. Uber will not allow you. Of course they will say, "for security reasons".

In the current context, people with the skill, drivers, have no say whatsoever when it comes to their activity with any business partner, while the business partner can treat them however they want. Within the new business model Uber is promoting dynamic upfront pricing. This means that what you see is what you pay, unless there is an exceptional situation when the price can go up or down. Many times though, if we find an alternative shorter and faster route than the one seen by Uber as the most efficient one, if we manage to save, let's say, five miles, at the end we get paid less than what we were shown when accepting the trip. Uber wants us to do the miles, no matter what. And we don't know if they lowered the final fare for the rider, or if they left it the same and kept more from the fare at our expense. These are details we can never know, and in Ubers' view we should not know. If we complain they will flag our accounts as fraudulent, meaning that we're trying to extortionate Uber and the riders.

Since we have joined Uber, they've expressed a highly racist behaviour, organising parties and events catered only for Muslim and Indian celebrations and tastes. Not once did Uber wish us Happy Easter or Merry Christmas, while at Diwali and Ramadan we always received messages in the Uber Driver app. We are not saying it's wrong to wish them well during their holy days, but as a corporation working with people from different backgrounds, shouldn't they include every race and religion? Eventually this will come and haunt Uber on the long run, as it is unacceptable, racist, and discriminatory with clear intent. We come from a white Christian country and civilisation older than the Greeks that never attacked anyone, but which was attacked by migratory Arab, Slavic, Turkish and Hunnic people, etc., that killed millions of Dacians/Romanians during millennia, so how should we feel when we see this type of racism, with Orthodox Christians being left on the outside, and forced to celebrate only holy days from other cultures, with some form cultures that decimated Dacians (Romanians) even till modern days? We are celebrating all cultures and religions, but what is wrong with celebrating the Orthodox minority? There are plenty of Dacians (Romanians) on the Uber platform. There are about 300 million Orthodox christians in the world, 1.15 billion Hindus, and 1.9 billion Muslims. If Uber is promoting inclusion, why are they avoiding another amazing culture, the Orthodox Christians, a minority in the world? What's wrong with Hindus and Muslims celebrating our holy days too? We would love to see Uber catering to our culture, with *papanasi* and *cozonaci*. This would have never been emphasised if all cultures would've been celebrated, including Orthodox Christianity. We love celebrating holy days from any religion and with any culture because we are good people, and because bad history must be forgiven and not repeated, but when companies like Uber practice such racism, they are only creating gaps between races and cultures on their platform, bringing back bad memories in the minds of people like us.

When we are being forced to celebrate certain cultures while being discriminated against it can only make us go back to a defensive mode, unwilling to participate, despite having had celebrated Diwali and Ramadan in the past (and we still do) with Indian and Arab friends. They are all beautiful holidays that must be celebrated together, not separately or in a discriminatory fashion.

3. Rumours From Canary Wharf Bankers

The case against Uber in London came as a surprise to many, especially since other ride-sharing platforms were still operating in the city off the radar, with way worse service. According to some bankers from Canary Wharf, there were rumours that someone from Uber's previous management had slept with the wife of a British MP, and when the MP found out, he vowed to destroy Uber in London. While the bankers couldn't provide any concrete evidence, they believe that there may be some truth to the rumours. As we stated before, Uber drivers talk to many important people. We even had people that worked with Boris Johnson during his mandate as Mayor of London. That's a whole story for another novel, confessed during a half an hour trip. Along the way, we discovered that all these conspirators do have some good people within their ranks that want to fight the system from the inside. We even found out that there are Freemasons in the UK that have become Freemasons just to try to put an end to the satanic cult from the inside. Hearing that from some of their members, without opening the subject yourself, does make you think that there may yet be a chance for our salvation.

The fact that Uber was considered to be really bad overnight, while millions were enjoying the service in London alone, makes us believe that there must have been something powerful in the middle to trigger such an uproar. As you will find out later on, everything presented to you in the media was complete misinformation, nonsense, and released by so called trustworthy authorities. Because of one driver not protecting his login details, all other Uber drivers were portrayed as rapists. If so, who was completing the millions of trips each month, and why was the user base getting bigger?

Whether or not these rumours are true, the fact remains that Uber faced a significant setback when Transport for London (TfL) decided not to renew its operating license in 2017. This decision was met with outrage by Uber's millions of London-based users, as well as the company's drivers, who relied on the platform for their livelihoods. However, TfL cited "a lack of corporate responsibility" as the reason for not renewing Uber's license, citing concerns over the company's approach to reporting criminal offences and background checks on drivers. TfL also expressed concern over the use of Uber's "Greyball" tool, which was designed to evade government regulators. We don't know anything about this practice, but as driver I know that I could not do anything illegal on the app, to avoid authorities or not, as Uber has always recruited new drivers based on them having a licence issued by Transport for London. The only ones to blame for driver checks was TfL! And all along you were being told that Uber was not doing proper checks on their drivers. We really hope this settles this misunderstanding once and for all. We cannot believe that the media succeeded to make the public think that Uber was doing the checks. This is how you see a pattern again, of political intervention and media manipulation. All along TfL was to blame and not Uber. No driver could join Uber without a valid TfL licence.

In response, Uber launched an appeal and was granted a 15-month probationary license in June 2018. During this time, the company was required to address TfL's concerns and demonstrate that it was fit to hold a full operating license. This again is madness to think that Uber was not fit for purpose, when other platforms, like Bolt, were going rogue in the market. Go on and research the Taxify (Bolt) fraud case. They rebranded as Bolt and TfL reissued them a licence as if nothing ever happened. No fuss, no court case. This reinforces what the bankers have told us. And Bolt is a very unsafe platform, we'll tell you that, from experience.

Despite Uber's efforts, TfL announced in November 2019 that it was once again refusing to renew Uber's license, stating that the company was "not a fit and proper operator." This decision was upheld by a judge in June 2020, and Uber was forced to appeal once again. In the meantime, Uber has continued to operate in London under its probationary license, which was extended until September 2021. The company has made significant changes to its operations in an effort to demonstrate its commitment to safety and compliance, including the implementation of a 24/7 incident support hotline and the introduction of a "Real-Time ID Check" feature, which uses facial recognition technology to ensure that drivers are who they say they are. All other competing companies did not have half of these features but they were free to operate.

Uber's reputation has been tarnished by a series of controversies, including allegations of sexual harassment and discrimination, as well as a widespread backlash over its treatment of drivers. We'll talk about the latter in detail, about what actually crashed the service to what it is today, in 2023.

In the end, the case against Uber in London may come down to more than just allegations of corporate wrongdoing. Uber changed so many managers that eventually the one desired by the regulators fell into place. It could also be a reflection of the broader cultural and societal tensions that have surrounded the rise of the gig economy, as well as the ongoing debate over the role of regulation in the digital age, what is actually holding it back from evolving.

One of the main arguments made by Uber and its supporters is that the company provides a valuable service to consumers, which is absolutely true, offering a convenient and affordable alternative to traditional taxi services. Uber has also argued that it creates economic opportunities for drivers (which were wiped out in March 2022), many of whom are able to work flexible hours and earn extra income through the platform. That income is now barely survival or less.

Uber's business model has also faced criticism, with many arguing that the company exploits its drivers and fails to provide them with basic labor protections. There have been numerous cases of drivers suing Uber over issues such as pay, benefits, and working conditions. In addition, Uber has faced a huge backlash over its data privacy practices, with some arguing that the company's use of personal data is invasive and raises significant privacy concerns. The company has also faced allegations of discrimination, with some riders and drivers claiming that they have been the victims of racial or gender-based discrimination while using the platform. We are not saying that some of these allegations might not be true but each case needs to be isolated and assessed individually, and not made a generalisation. Drivers face exponentially more abuse from riders than the other way around. You cannot even compare the numbers actually. And Uber drivers are nowadays highly professional and experienced. Drivers asking for working rights from Uber is one of the most outrageous requests coming from a self-employed individual. It's what the regulators and conspirators have used against the gig economy, to destroy it.

Another piece of interesting news we got from people high up in London's financial district is that a certain company competing with Uber is used for money laundering by a certain criminal network. We won't press this button anymore, for obvious reasons. What we do know is that this type of rumour travels with *lightning* speed... Don't bother with it, most probably it's not true.

4. Why it changed

In London, a group of Uber and other private hire drivers formed a union called ADCU (App Drivers & Couriers Union). In our opinion, forming a union as self-employed is an aberration, as YOU have decided to become self-employed. No one forced you. Because if you form a union, you'd have to ask each individual acting in the same market if they want the changes, and if they don't, their will be done too. These drivers (the Plaintiffs) were unhappy with their pay and working conditions, and decided to take action by stopping their driving activity and filing a lawsuit against Uber. No one knows where they got so much money from, to file for the lawsuit using top lawyers in London. Also, no one knows the actual number of drivers they were representing. From their public communications it seemed as if they were representing everybody, which is totally misleading and hurtful.

The lawsuit was filed under the names of the ADCU founders, followed by "and others".

The drivers' main demands were for a minimum hourly wage, holiday pay, and a pension. They believed that Uber was not offering them these basic employment rights, and that they were being treated unfairly. To support their cause, the ADCU began recruiting more drivers to join their ranks and protest against Uber. Uber drivers were not and are not employees. We are all self-employed! A taxi/private hire driver can never be an employee! It is the same thing as creatives on Fiverr or Freelancer. You cannot ask for benefits from the entities you decide to offer your service through, when deciding to be your own boss. Destroying all others' business on the platform for your own interest is a breach of individual rights, and it should be applied only to the one requesting the changes. All other drivers (the majority actually) have been stripped of their legal rights. They had no legal representation, and that was unfair. What the unions were asking for, what the court decided, and what Uber accepted is equal to crime. Uber drivers are not employed by Uber and they can never be. The ADCU probably had the backing of other players in the transport industry and politicians. This support may have given the drivers more unnatural confidence and motivation to continue their campaign against Uber. Maybe important financial support too?

The ADCU, more specifically the union's founders, Yaseen Aslam, James Farrar, and Robert Dawson, "and others", eventually took Uber to court, and the case was heard by a judge. The plaintiffs argued that they were entitled to the same employment rights as other workers, and that Uber was exploiting them by not providing these rights. After a long and complicated legal battle, ADCU finally won their case. However, none of the two main plaintiffs ever returned to work for Uber after winning the case. In fact, one of them even secured a contract with the British MOD (Ministry of Defense) with his other IT venture. Yaseen Aslam was later the subject of a paid Forbes article portraying him as a hero in London for managing to bring Uber to its knees. We've seen this type of paid content before… Forbes will never write a positive article about anybody if not paid or at least instructed to do so by rich powerful influencers. And why praising only Yaseen Aslam and not James Farrar too? And how come Yaseen Aslam got a contract with the MOD straight after that? How did he earn his money without driving for Uber, during the years of legal battle 2016-2021, if he was struggling as Uber driver? Who were the "others" in the case, and what was their number? There are people of Aslam's culture who've said that they called him a traitor on the street and spat at him for what he's done. We would like to see a Forbes article writing about those incidents. Can these two guys be undercover agents? Based on the above they cannot not be, as there are so many things that don't add up about how they managed to survive throughout all these years and who sponsored them.

After winning the lawsuit, ADCU still has an online presence, but legally it was dissolved. Anyone can check this on the Companies House website. The lawsuit allowed them to get registered as a legal union. Another union that supported the cause is GMB. We tried to contact both unions for more details about their members but they both refused to reply, as if the number of union member is a State secret. This is just adding to what this book is all about, that a few minds with criminal intentions have ruined a working industry. Most drivers we spoke to, even at Uber's Greenlight Hub, were not members of any taxi and private hire union.

In hindsight, why in the name of God would you ask for workers' rights if you're not a worker? A worker is someone on PAYE! As long as you are filing your own taxes, you cannot be a worker. The unions, Uber, TfL, the Mayor of London, the Supreme Court knew of the consequences of such a decision. They also knew that most drivers were not asking for such rights, having their rights breached by a shady minority driven by malicious intentions. How can you ask to be paid minimum per economy if you're earning £5,000 per month only by doing UberX? Why would you ask for holiday pay when you're earning £5,000 per month? Why would you ask for pension when you can afford to invest in your future at least £1,000 per month that you have freely in your pocket? And also, when you can go offline for as many weekends as you please? Keep in mind that this was during times before the pandemic, when living costs were way cheaper. The real issues with Uber were different and we'll talk about them later on. What's even more worrying is that there is no one out there to talk to about your situation. TfL is blind and deaf, unions are not replying, the Supreme Court cannot be moved by your despair, and Uber turned into just another malicious corporation of the global "green" agenda, i.e World Economic Forum, a terrorist organisation by any definition. A green agenda that is killing people more than saving. Say hello to the new Middle Ages of 2023!

5. Regulators

In the bustling metropolis of London, the regulator and licensing authority for private hire transportation is none other than Transport for London. This esteemed organization is responsible for issuing private hire driver licences and ensuring that all drivers meet the minimum requirements for safety. However, it is important to note that the role of Transport for London is not to over regulate the industry, as this can have dire consequences for both drivers and riders.

Over regulating the private hire industry, by making excessive changes to the basic rules and regulations, can put undue strain on the drivers and riders who rely on this mode of transportation. It is important to remember that while certain changes may benefit a minority, they should not be implemented at the expense of the entire industry. Instead, the minority should only have the right to request that any changes only apply to them and not to all drivers, if they are not unanimously agreed upon.

It is also worth noting that for all the wrongdoings that Uber has been accused of by Transport for London, The Media, and the Mayor of London, it is actually the responsibility of Transport for London to conduct security checks on drivers and ensure their compliance with regulations. Despite this, Uber has often been made the scapegoat for any issues that arose, when in reality they are only responsible for asking new drivers to prove that they are licensed by Transport for London.

In fact, it is not only riders who are at risk in this industry, but drivers as well. The chances of encountering a dangerous rider are exponentially higher than those of a dangerous driver, yet Transport for London does little to protect the drivers who are out on the roads every day. There are no checks on riders outside of the ratings set by drivers, leaving drivers vulnerable to harm. It's not as if TfL could check riders, but they could actively educate them on how to behave through their different channels. There was absolutely no campaign from the regulator aimed at educating riders on how to use other means of transportation available licensed by them, other than the TfL network. Uber could do more about this too, and this is what TfL should've enforced.

It is also worth noting that Transport for London has always asked more from Uber than from competing platforms, which raises questions about the fairness and impartiality of their actions. Even the introduction of the congestion charge for hybrid private hire vehicles is discriminatory, as there are new black cabs that are hybrids and do not have to pay it.

As regulator, Transport for London should be impartial and unbiased in their actions. However, it appears that this is not the case, as their sudden and dramatic change of stance towards Uber remains unexplained. This only serves to reinforce suspicions of bias, particularly in light of the rumours circulating in Canary Wharf.

Overall, while it is important to have regulators in place to ensure compliance with safety standards, it is crucial that they do not over regulate the industry and put undue strains on drivers and riders. Transport for London, as the licensing authority for private hire in London, must strive to be impartial and fair in their actions towards all players in the industry. It is time for Transport for London to come clean on why their view on Uber changed so dramatically overnight and what triggered it, in order to restore trust and fairness in the industry, and most importantly in the regulator.

The wheels of Transport for London's accusations and tactics have ground the private hire to a halt, leaving millions in shambles. The drivers, the lifeblood of these ride share companies, have been left stranded and helpless. The relicensing conditions imposed by Transport for London have been a constant source of frustration, as drivers are left with no means of communication or recourse.

The worker status, so eagerly sought after by an uneducated minority together with TfL, has reduced all drivers to mere cogs in the machine. The power and autonomy that once defined their profession has been stripped away, leaving them at the mercy of the very companies that once promised them a brighter future.

Despite the billions of pounds that these ride share companies have pumped into the economy, the drivers are left to bear the burden of Transport for London's accusations and tactics against the industry. They are left to struggle financially, with no one to turn to, and no one sane and impartial to speak on each driver's behalf. There has to be a system in place to give each driver a voice.

The once bustling streets of London have turned quiet, as the private hire industry struggles to keep pace with the relentless demands of Transport for London. The drivers, once the backbone of this thriving industry, are now left to navigate the treacherous waters of relicensing conditions and worker status, with no one to support them. This can only have dire consequence on drivers' and riders' health and wellbeing.

The accusations and tactics employed by Transport for London have left a gaping hole in the private hire industry, and it remains to be seen if it can ever fully recover. The drivers, who once played a vital role in the economy, are now left to pick up the pieces and move forward, with nothing but their determination and questionable resolve to sustain them.

What TfL fought for with Uber for years, which led to the sudden cancellation of their licence in London, was that the contract was between the driver and the rider, when they said it should be between Uber and the rider, for tax purposes. But that made no sense to ask for as most of the income on the Uber platform was going back into the real economy, which is exponentially better for the economy than paying corporation tax on less than 25%. So, less money was being spilled outside of the UK than with local taxi companies owned mostly by immigrants that are sending a lot of money outside of the UK, mostly to Asian countries. TfL eventually got what they wanted at the expense of driver financial security. Now, Uber is not showing drivers how much the rider paid, although the driver has absolute power over the contract between Uber and the rider. Without the driver's acceptance, the contract cannot start or be executed. So, how can it be a good thing to keep the driver outside of this contract? Legally and technically it makes no reasonable sense to push for such a change. The current Mayor of London has a huge responsibility, and for what he has pushed against Uber, he's got a lot of explanations and compensation to give to drivers on this platform. Besides the fact that he is renowned as not being very sharp and as the most corrupt mayor London has ever had, the man that took the fun away from this amazing city is obeying the imposed "green" agenda by the letter, no matter the consequences. With this type of people running things, it is impossible to have a bright future. It'll take a long time to resolve the mess he's made in the city.

As the private hire industry continues to grapple with the harsh realities of Transport for London's accusations and tactics, one can't help but wonder about the true cost of this ongoing battle. The drivers, who have given so much to this industry, are now left to bear the brunt of its consequences. The future of the private hire industry hangs in the balance, as the drivers struggle to find their footing in the face of these uncertain times.

6. The Last Two Big "Secret" Cult Meetings In The UK, Plus A Tercentenary One

This point might seem outside of the scope of this book for the reader, but we can assure you, it is not. In order to understand a situation exhaustively, you must be able to see the whole picture, with little focus on smaller updates. Actually, small situations are meant to distract you from the big picture.

As we said before, Uber drivers speak with many people from all backgrounds, including politicians, bankers, journalists, etc. Around

2018-2019 we started driving freemasons to this, apparently, outstanding freemasonry meeting, to lodges across London. They were heading there from all directions. It was unusual to see so many people dressed roughly the same, and joining what seemed as an AlphaPi-like American brotherhood. Jokes aside, it was the first time we had direct contact with people from an apparent secret organisation. Obviously they are not so secret. We remember wearing t-shirts, so it must've been the summer of 2018 or 2019. Prior to that, on 31 October 2017, freemasons from around the world gathered at Royal Albert Hall to celebrate 300 years of activity. We didn't give it much thought before talking to other people, who've told us that when the freemasons are having meetings, something big is usually going to happen in society sooner than later. And that meeting was big, we can tell you that. We connected the dots late into the pandemic that started roughly in August 2019, when the inventor of the PCR method, Kary Mullis, one of the most remarkable scientists the world has even seen, a Nobel Prize laureate, died suddenly of pneumonia without any mention in the mainstream media. And to give you more insight into how severe the pandemic was, no Uber driver passed away from this new virus. Taxi/private hire drivers are the most exposed groups of people out there, even more than doctors and nurses (confirmed by them too), which are first line workers. Drivers were driving ill people and doctors to London hospitals. Same places where patients were dying under medical supervision, only due to the new coronavirus.

So, other people were right. Whenever there is a freemasons meeting, something big is coming. May God help us all and bring peace on earth, we say!

Now, in order to understand a conspiracy, and stop acting all crazy when you hear this term, you need to be able to make the difference between a "conspiracy" and a "conspiracy theory". First of all, any theory must be disproven, not proven. You prove a theory by not being able to disprove it. When someone is serving you a theory, you, the second party, the receiver, must start doing research and try to disprove the new information. This is what all scientists do to each other's views. Most people receiving the new theory start acting all Karen (as per a trend in the US of A), and ask that the one that comes up with the theory brings proof of their assertion, which is the completely wrong way of how things should evolve. Plus, if that was the case, the one with the theory would be exposed to ridicule, while doing all the work himself on top of discovering the theory in the first place. The receiver must engage immediately in disproving the new theory. Once the theory cannot be disproven, it then becomes a fully fledged conspiracy, meaning a REALITY you should accept! A conspiracy is something ongoing, true events. This is the difference between the two: one is real, and the other yet to be disproven, and NOT proven by the initiator. You prove a theory by failing at disproving it!

Coming back to the topic, around May-June

2022 there was another freemasonry meeting. This time we had freemasons disclosing their statute and that they had enough, that some are fighting from within their ranks to destroy the cult. We didn't really believe any of that, as we do not think that is possible. We know too much about what they do to trust them with anything, especially words. This time we were more aware about their meeting and the outcomes. We knew something big was coming again, but we didn't know what. And it did. In September Queen Elisabeth II died. And it happened in such a brief and shady way that it turned everyone blind. No body was shown to the people, no last recorded message from the longest reigning monarch in UK's history, not even a video with her receiving and talking to the new Prime Minister, as she usually did. Just a photo with them stretching their hands toward each other, but never actually touching palms. How can the longest reigning monarch of the UK not leave a last message to her people? And all after inaugurating the Elizabeth line in London a couple of months back, looking strong, happy, and quite healthy. We guess that was the big event following the freemasons meeting. So our riders were right. Weirdly enough, after that, we found a book, published in 2010, which stated that the Queen will die in 2022 at 96 years old…

How is the above relevant to this book? Well, in the last three years we have witnessed the intentional crash of worldwide economies, and governments wanting people to get very poor, while promising state financial aid which they knew will get people receiving it even poorer. Because this is how the economy works. The Supreme Court's decision following a legal battle started by a minority of Uber drivers came in 2021, and it was implemented by Uber, voluntarily (nothing could force them to abide by it), in March 2022, straight at the end of the pandemic. First of all, TfL, the Supreme Court, the unions, and everyone else involved knew that the only effect this implementation will have was the financial crippling of drivers and riders, on top of making the system highly unreliable. Uber was never going to pay those benefits from their 25% share. As consequence, Uber is making very big money, while stealing it from drivers and riders. And riders don't know what's going on, why the system they're all relying on has now become at many times unusable. This is clear evidence of a large conspiracy aimed at destabilising millions of people, because drivers not earning enough money is having the desired effect on the overall economy – its intentional crash. Even that guy from the World Economic Forum, Klaus Schwab (a terrorist by any definition) has stated in his disgusting books that he wants to take the GDPs of all countries to MINUS, thinking it is not a healthy metric for a country's wellbeing in the new world, or the "green" agenda. If this happens, hundreds of millions of people will die prematurely. They will be literally killed!

And this is how Uber has been taken over by entities of the above agenda, implementing abusive policies that are meant to keep drivers on the edge of survival, riders stuck or paying a fortune, and its shareholders very rich. This is the bigger picture within an even bigger picture, of poorer people having to work even more for way less. From a company that has improved millions of lives in the UK, Uber is now destroying even more millions, with clear intention, a fully fledged conspiracy. It all coincided with the naming of a former British Airways executive to run Uber in the UK, Andrew Brem. During his time at BA, the airline became identical to low cost alternatives, while charging premium prices for their tickets. BA stopped giving snacks and drinks for free around the time he was named CCO (Chief Commercial Officer). Wherever this guy goes, things get worse for everybody, while the shareholders get fatter.

7. The Green Agenda (Agenda 21 From Rio, etc.)

If you've never heard about this before, just go ahead and purchase the UN's Agenda 21 from Rio. It's the constitution by which almost all governments must abide in order to build a "sustainable" approved future. It does not include your personal view and plan. You'll learn that everything happening to you is not random, but set in stone decades in advance. It involves the deindustrialisation of nations and a throwback to a new type of middle ages in which you have no say. They might give you the impression that you do, but you don't. Whatever they decided goes, no matter what and how you vote. They're on either side to pick it up. And they're naming their officials in key positions of power, while undercover agents and other types of conspirators will take positions in strategic corporations. One of these corporations is Uber, as they are involved with building the future. And building the future within the Agenda 21 is not possible with everyone being financially secured. It is a well known fact that black cab drivers in London make somewhere between £3,000-£5,000 per week, more or less, while an Uber driver could make in 2019 about £5,000 per month, in up to six days a week. For the green agenda to go forward it had to get the money out of as many hands as possible, and because taxi drivers have MPs representing their interest, private hire drivers had to fall. Some cabbies are freemasons, conspirators, informers, etc. Too many millions of people were living good lives and going about too easily. With their income heavily reduced (50%-60%) overnight, Uber drivers have to stay on the platform for seven days a week, work very long hours, reject thousands of trip requests a month, and be forced into switching to the highly inefficient electric vehicles without any financial support, all while black cab drivers received a £50,000-£60,000 non-refundable grant from the government to purchase new cabs. And many new black cabs are not full electric, but hybrids. Yes, only Uber drivers must go full electric.

    Just imagine the horror of Uber drivers today when they're struggling to earn around £3,000-£3,500 per month, they have to pay around £1,200+ per month electric vehicle rental, about £1,000-£1,200 to charge the vehicle for 30 days, excess of about £500-£1,000 if an accident occurs, and other costs. Drivers also lose money while having to wait for the vehicle to charge, on top of everything. Having to reject trips because there are not enough miles left in the battery is highly frustrating and adding up to this nightmare. This is enough to showcase their tragedy and why the "green" agenda means destroying the lives of millions. Drivers are self-employed, not workers, and no one from TfL, Uber, or the Mayor's office is willing to listen to them, while these illegitimate unions are making it worse, while pretending to help. And all this with drivers not being able to book time off anymore. It's constant grind on the roads. If that is not a health & safety concern for everybody, we don't know what is. All created by the beautiful authorities that were meant to stop something like this from ever happening to the industry and people. It's never a case of them having made a mistake, they have specialised studies about the repercussions of any of their actions. They knew all along, and also that it was a minority of people asking for these changes, who the people behind the legal actions are, why everything started, how much money we were making before, they knew everything.

The "green" agenda is one of the most disruptive and destructive agendas in human history, that is being actively and aggressively implemented no matter what you say. It is just a switch from one energy source to another, with increased toxicity. Calling electric vehicles running on batteries "green" is an aberration. Calling for a decrease in CO2 emissions worldwide when we are at a historically very low level of atmospheric CO2 compared with the oxygen levels is nothing more than a criminal intention. It is proven that electric vehicles on batteries are maybe ten times more dangerous for the planet than full fossil fuel ones. But we are being managed like sheep toward an even more centralised way of living, which is actually the problem. Centralisation is humanity's destruction, hence the problems with the Uber platform. A system that used to work is not working anymore because of over regulation. There is a business model of Uber that will work, and that is the previous model, before upfront pricing and treating drivers as workers.

"Green" is the opposite of wellbeing, safe is the opposite of healthy, centralisation is the opposite of freedom, overregulation is the opposite of protection. Welcome to dystopian London!

8. The Mayor Of London

It is a commonly held belief among many Londoners that Sadiq Khan, the current mayor of London, will go down in history as one of the worst and most corrupt mayors the city has ever seen. While his tenure has been marked by a number of notable achievements, such as the introduction of the "hopper" bus fare and the expansion of the city's bike-sharing program, many Londoners feel that his legacy will ultimately be defined by his heavy-handed approach to governance, which has been characterised by a seemingly endless stream of taxes, restrictions, and regulations. Talking a bit about his views on cycling around London, how much can you cycle before becoming a pickle? He wants everyone to cycle for hundreds of miles, from bankers and bourgeoisie, to bin men and the disabled. Cycle, damn it! He does not want drivers to drive anymore, even if you switch to the highly damaging electric. He went so far into the absurd that he thinks electric bicycles are "green". No, Sadiq Khan, classic bicycles are green. His policies and views are always shaped to cocoon and accommodate the big corporations at the expense of Londoners' finances and wellbeing. If this continues London will become a living hell very soon. He is also supporting and pushing forward a charge to drive into London, on top of the daily congestion charge, leaving millions of people in shock.

One of the most controversial aspects of Khan's tenure as mayor has been his relentless pursuit of installing speed humps and other traffic-calming measures throughout the city. While these measures may have been intended to make the streets safer for pedestrians and cyclists, they have instead caused significant damage to vehicles and led to increased travel times for emergency services. This translates into thousands of people dying needlessly. The risks of installing speed humps heavily outweigh the benefits. This has led many Londoners to question the mayor's priorities and his understanding of the impact that his policies are having on the city.

Another major point of contention for many Londoners has been Khan's stance on Uber and other ride-hailing services. Despite the fact that these services have been hugely popular with Londoners and have provided a much-needed alternative to the city's often unreliable and overpriced taxi services, Khan has been a vocal opponent of Uber and has lobbied against the company at every opportunity. This has led many to accuse the mayor of being out of touch with the needs and desires of the city's residents and of prioritising the interests of powerful lobby groups over the well-being of ordinary Londoners.

Starting 01 January 2023, Transport for London is no longer licensing hybrids for private hire, but only zero emissions capable (electric, and plug-in hybrids). And it only applies to private hire. Hybrid black cabs are no danger to London's air. This policy is Sadiq's doing, pushed forward at the only time in history when private hire drivers are struggling to survive. In the context of drivers earning 50%-60% less compared to 2019, how can such a decision be taken without consistent financial support? Do private hire drivers' lives and their families' not matter, even after being the heroes of the pandemic when taxi drivers were hiding? Where are they supposed to get £60,000 from in order to buy an electric vehicle with reliable mileage per charge? And if we would have £60,000 lying around, why would we invest it in driving with Uber, when the corporation turned against us?

At the end of the day, the legacy of Sadiq Khan as mayor of London will be a mixed one. While he has certainly accomplished some notable things during his tenure, such as the introduction of new transport services, his heavy-handed approach to governance and his lack of respect for the working people has left many Londoners feeling disappointed and disillusioned. In order to truly be a successful mayor, it is essential that a leader has integrity and respect towards the working people and understands the impact of their policies on the city and its residents. Unfortunately, Mayor Khan falls short of that standard.

He will be known in history as the mayor London did not need.

9. Do No Harm As Company

There are times of plenty, and times of crisis. We are living the latter, it's a sad fact, but there is hope. No, better yet, there is belief, as we think that people that hope are waiting for someone else to do something, until nobody does anything. People who believe on the other hand, these are the people you want to be surrounded by. These are the ones that make things happen. And at times of crisis they're coming out of hibernation. When the world is being put on its knees, that's the pivot that will show you the true men and women, and also whoever is just hoping, faking heroism.

Corporations have been rogue for too long, and politicians corrupt for long enough. We know it's not about money for them, as they have more than they could spend in 100 lives. It is about power and control over the hopefuls and believers. But it needn't be like this. These trying times are the perfect opportunity to take control of your life because everyone is vulnerable, including corporations and politicians. In times of crisis nobody is safe! Keep this with you your entire life. We're not talking here about hurting each other, but about taking the right legal actions to free yourself more, opposing the increased push for enslavement. Their plan won't succeed and they know it. You need to know it too. That is why it is time for corporations, old and newly incorporated, to sign and abide by a "Do No Harm To Humans" affidavit. They either do it voluntarily or are being forced to dissolve immediately if they refuse. This is what you should ask from the corporations you're dealing with. We believe that human well-being should be prioritised over saving nature, because if people are not thriving, nature will get destroyed by default. On the other hand, the destruction of the planet is a direct consequence of people management, instead of helping and decentralising the individual.

Starting 2022, Uber is hurting its riders, drivers, and we're pretty sure it is valid for their employees too, as they never looked happy whenever visiting the Greenlight Hub, or as if they'd earn a fair wage.

The problem with what Uber is doing to everybody is that they're destabilising the economy, and implicitly hurting humans.

No company, private or public, should serve an agenda outside of the initial purpose of the business model, and especially when it is clear that overregulation of the sector will have as only repercussion the crippling of the business model, it's beneficiaries, it's business partners, the economy, and the innovation altogether. When a company is aware of a severe degradation of its business model and it's customers' experience coming via regulatory channels, even if investors would gain even more from these changes, the company has a moral duty to oppose such changes at the cost of the entire business. And most of all, drivers are/should stakeholders in Uber. No business should do harm. Business is intended to ease human condition, to constantly improve it. If the initial business model worked for everyone, the business cannot later change to comply overregulation that would only benefit investors, while everybody else loses a quality service that made their lives way easier. A true company would fight these agendas, investors, regulators, until finding new ways to improve the services offered, and not to degrade them. Uber is today complying with external agendas pushed by highly influential entities at the cost of the entire service.

The concept of "doing no harm to humans" is one that is becoming increasingly important in today's business world, and regulators are not making it easy to achieve. As consumers become more socially and environmentally conscious, they are demanding that companies take a more responsible approach to their operations and ensure that they are not causing harm to people or the planet. The push for electrification is a dangerous one, as it is not green. Before the changes to private hire Uber has been vilified for creating congestion in central London and polluting the air. In 2023, London is more congested than ever before, with an exponentially lower number of Uber vehicles working there. So, were Uber drivers contributing to the congestion? Definitely not. If they were, then cabbies were too. Were Uber drivers polluting the air in London? Absolutely not! The majority of drivers were driving hybrids that switched to electric mode most times while caught in central London traffic. All while many cabbies still have no restrictions on diesels and hybrids. Why else would they push for all these restrictions on private hire drivers working with Uber in particular, if it didn't have political underwriting? Everybody, from the Prime Minister down, knows what is happening with Uber riders and drivers, and probably they are laughing. Will you accept this state of affairs? Let's say you're a stranded lady in distress, with no other means of transportation available than booking an Uber. You open the app and tap "confirm booking", but no driver is picking you up despite seeing cars around you. You wait for half and hour, one hour, over an hour, but nobody comes. What do you do? We had this situation endless times, when women and mothers were stranded, and desperate to leave an area or just go home. This is what Uber is for you – comfortable and affordable transportation anytime, anywhere. The "where to, love?" guys would never work in that area, they do only central London. And if they'd do, they would charge you 3-5 times the Uber price, because you are in distress, therefore willing to pay any amount. "Can you pay cash, love? The card machine is broken…"

In conclusion, the concept of "doing no harm to humans" is one that is becoming increasingly important in today's business world, but it is being manipulated in the wrong direction. Uber is bound to become very dangerous for the environment and humans once the switch to full electric is complete. There will be more dangerous particulate matter in the air for you to breathe, due to the weight of these electric vehicles (about 400kg extra compared to a Toyota Corolla). And if you didn't know already, the dangerous chemicals footprint of an electric vehicle will never be offset during the whole life of the vehicle, because the battery will have to be changed at least once, as no electric vehicle will ever pass the 500,000 miles threshold in order to even achieve 90% toxicity offsetting. And all along the solution was there, without financial burden for drivers. Old hybrid vehicles, and even diesels, could have been made zero emissions with a hydrogen-on-demand device fitted to the engine, with zero risk of explosion as no hydrogen would have been stored in the tank. This is very old technology that is not accepted by regulators, for obvious reasons. Their motto is (actually WEF's motto): we need to switch to electric, no matter the harm done. It is for this reason that a "do no harm" policy must be pushed by the people and steered in the right direction, of improving people's lives, outside of politics and corporate manipulation. We believe this can be achieved using decentralised blockchain technology, maybe with a DAO (Decentralised Autonomous Organisation).

10. Going Back To The Middle Ages

This can't be familiar to anyone living today, but doesn't it feel this way? We had the Concorde airplane operating commercial flights in the nineties, and in 2023 this seems unachievable. Cola and drinking water were being commercialised in glass bottles. Today everything is wrapped in plastic. And so on. In the 1900s you had to thank for having running water and electricity, in 2023 you can't afford to pay these basic utilities, while being told to use less or no water. Between 2012 and 2020 Uber has improved the lives of millions across the UK, but in 2022 the system stopped working, making it a nightmare for all that now depend on this service. Instead of increasing the speed limit and getting rid of all road obstructions, including the speed humps, and educating pedestrians and cyclists that they should not jump in front of a moving vehicle (as it happens to us very often), we're restricting everybody to block London. For a few bad drivers that should not be on the roads and irresponsible riders and pedestrians, everybody needs to suffer. Instead of forcing companies to pay people more, we must accept higher taxes and prices for less money, to save Ukraine, or the penguin at the south pole. Although we need to relax and get rid of passports and borders, we have to accept increasingly stringent checks when travelling. Nowadays you are a terrorist first, and then not guilty if bringing proof.

More passports, more medicine, more diseases, more politics, less children, more charities, less solutions, sealed patents, more tanks, more secrets, more dead people, less planet, less movement, less flights, dirtier air, more electric vehicles, no free speech, more criminals, more pardoned pedophiles, more polarisation, more conflicts, less solutions, less God, more religions, no exit, no entry, barbed wire, potholes, no compensation, less meat, more secret societies, more McDonald's, less sugar, more plastic, less animals, more vegans, more rich people, more poor people, more guns, more wars, less peace, horror movies, be afraid, the sun is cancer, the seas are poisoned, don't move, you'll die, welcome The Fourth Industrial Revolution (The Fourth Reich), don't breathe!

Coming back to Uber, why are we going backward if this system used to work? We are not the only ones thinking that we are being driven off the cliff of existence prematurely, as many others smarter than us have said it decades ago, but what you must understand is that all is not lost. Good people need to stand tall and take individual action to oppose what's coming. We've had hundreds of riders complaining about not being able to use the Uber platform anymore due to very long wait times to get a driver despite paying a lot, and in their vast majority, they were not willing to do anything. They didn't know that Uber is charging them more and paying drivers way less, and even after explaining the situation, many didn't show any empathy for us or irritability toward Uber's practices.

What passive riders don't understand is that if nothing is being done to protect drivers from the implementation of the Supreme Court's decision to treat drivers as workers (which they can never be), taxi drivers (who are considered self-employed/independent contractors) will rip off anyone, rich or poor. We already have complaints from riders that went to book taxis when Uber drivers did not accept their trip requests, and they were charging them at least 2-3 times the full price quoted to them on the Uber platform.

When we say that we're being sent back to the middle ages, we mean you getting poorer and the rich getting richer, while the economy is stiffer and less diverse, to encompass the criminal ideas of corporations like Google, Facebook and Microsoft, and organisations like the EU (it is no longer an image of its founding act), the UN, the WEF, NATO, etc. All this while your life is degrading at an accelerated rate. These are the best times to take individual legal action against these practices and organisations. You just have to say no to what they're proposing, and push it out there to block them. Quite a few councils outside of London have banned Uber from the beginning, to protect the local cab firms. Needless to say that the vast majority of Uber drivers are local, and the owners of those cab firms they're trying to protect are sending money to Asia, mostly. And so, many business people in these areas are complaining to us, when driving them home, about how in the mornings they have no private transport available to go to London. Since when councillors are dictating to residents about what services they should have or not? Residents were never asked, because most councillors are conspirators promoting the "green" agenda. Apparently one Uber driver driving a new Toyota hybrid (ULEZ exempt) is polluting the local air more than ten 20 years old diesel cabs.

We're going back very fast, and we don't like it! Are we going to accept it? Under no circumstance. Just remember that if you decide to sacrifice your life for future generations, you're sacrificing yourself for the children of the rich, who will continue the plan for destruction, and not for the good children of the poor. These ones will be working in the quarries of the future, dying while mining for Lithium or other rare minerals, used in technologies that Nikola Tesla wanted to avoid, like the inefficient solid batteries.

11. Regulators Turning Into

Oppressors

Transport for London is the voice of the Mayor of London. Every regulation is directly connected to his office. TfL is a corporation making about £40 billion per year, and a lot of this money is made without producing anything. The congestion charge is a massive contributor to this huge amount, probably after the train and bus tickets. From 2013 they reduced the personnel number significantly, replacing them with ticket machines, and even so, they have no money to pay higher wages to their current employees. TfL is the regulator for transportation in London. If the regulator is practising apparent managerial incompetence that's harming only the lower levels employees, how can we expect them to issue supportive regulations for private hire drivers? They could not wait to tackle Uber, the leader in the private hire industry. And since they couldn't find anything to pick on, as Uber was being managed really good before 2020 (despite an executive having slept with an MPs wife), TfL decided to cancel Uber's operator licence based on Uber not checking drivers properly. THIS HAS ALWAYS BEEN TFL's DUTY! Uber could not activate drivers without a licence issued by TfL. That licence means that TfL has done all background checks on the driver, and guarantees to Uber that the specific driver is trustworthy and no danger on the roads or for riders. They found the right moment when a driver that stole his brother's credentials raped a woman. From that moment on all

Uber drivers were rapists. Is a regulator supposed to entertain such hideous accusations? They had to, because they were to blame for any checks done. They had to throw the cat on someone else, and since most Londoners don't now how regulatory things work, then TfL must have been right, despite millions of trips being completed every day. How could the regulator take the blame? If people knew, they would've asked for a full restructuring of TfL. Even so, many voices think TfL should dissolve and be replaced with a modern, less centralised system. With this we agree. TfL is a waste of £40 billion that could build a truly modern and free London. Speak with any of their engineers, and they will explain how TfL is pumping the value of the contracts allocated to close companies, tenfold or twentyfold. Jeremy Clarkson had a row with them in the past about this irresponsible waste of money, which could be used to actually help people. Sadiq Kahn is smoking his shisha while all this is unfolding.

When TfL found out that they can cripple private hire drivers working with Uber by asking that the contract be set between Uber and the rider, and no longer between the driver and the rider, it probably felt as if they discovered the holy grail. Screw what the drivers and riders want or need! With this, TfL completely overlooked the real issues with Uber, that did not include Uber not paying the right amount of tax on less than 25% of the revenue on the platform. Some of the real issues that needed to be addressed were about better driver safety and better compensation for damages to the vehicle or when the rider didn't show up, etc. With the new business model way more tax is not getting paid, as money flows via different channels and companies, TfL, and Adecco, the pension manager, including Now Pensions. Everybody is making big money now, but not the drivers. And this is how drivers now get only 40%-60% of the fare, compared to 75% or more before the changes. If you were a driver, would you drive someone from A to B for as little as £0.30-£0.40 per mile?

While TfL was harassing Uber for all the wrong reasons, trying to mask their own incompetence and lack of responsibility, other competitors with Uber were doing business unrestricted, some being even relicensed after committing fraud, which Uber never did. Just research the "Taxify" scandal. Uber never had programs to avoid regulators either. It was a small group of drivers that cracked the code to make it possible to know where the riders were going. Before 2019, Uber drivers did not know where you were travelling to before picking you up. About 150 drivers have beed deactivated by Uber following the crackdown. But that was not a safety concern at any point. It was just driver adaptation to earning more than others. It was not fair and they got shut down. Journalists also told us that at least one ride sharing platform is being used to lauder money by organised crime, and most probably Sadiq Kahn knows all about that, considering him covering real issues by targeting only Uber, the standard in the transportation app industry. And so, the real issues with Uber were never tackled, but swept under the rug by a covert operation to denigrate the drivers and make it impossible for them to earn a decent living, despite working very long hours and actually saving lives every day. Their tactics worked, as they always do with corrupt politicians, and now nobody really wants to be or become a driver anymore.

12. The Supreme Court Involvement

The UK Supreme Court's 2021 ruling on the status of Uber drivers has been the subject of much debate and discussion. At the heart of this issue lies the question of whether or not drivers should be classified as workers, entitled to the same protections and benefits as employees. While a minority interested in making money (lawyers, unions, etc.) argue that this decision is a triumph for workers' rights, most self-employed from all industries believe that it is a misjudgment that ignores important factors and the will of individual drivers.

One of the most powerful arguments against the Supreme Court's ruling is that drivers are not employed by Uber, but rather are independent contractors who have chosen to be in partnership with the company on their own terms. Drivers are not and can never be employees of Uber. According to this view, the Employment Tribunal had no right to decide on the status of self-employed individuals, as it is the individual's choice to work as an independent contractor. The court should've dismissed the case as it never made the object of the purpose of an employment tribunal. Furthermore, the small minority of drivers who brought the case to court were seeking worker protections, while the vast majority of drivers were content with their status as independent contractors. Most drivers are not part of any union, and no one knows if any court asked for this essential piece of information during the proceedings, nor if they requested the results of the form sent by Uber to all drivers through which they were asking drivers about the desired legal status.

The biggest argument against the Supreme Court's decision is that it ignored the will of individual drivers. The court failed to take into account the fact that drivers had made a deliberate choice to work as independent contractors (and not be employed by a company), and that this choice was a result of their own individual circumstances and needs. By upholding the High Court ruling, the Supreme Court was effectively imposing its own views on drivers, rather than taking into account their individual needs and desires. A court of law is supposed to protect the well-being of people, and not to damage it. This is how you know that there must have been a political driving force behind these changes, and if so, the judiciary system in the UK will suffer a tremendous damage to its image.

A further point of contention is the question of

whether or not the Supreme Court should have called for a referendum of all drivers to determine their preferred status. This would have given drivers the opportunity to voice their opinions, express their views on the matter, and would have ensured that the decision was made in a democratic manner, taking into account the views of all drivers, rather than just a small minority. In the worst case scenario, the court should've mentioned in the decision that the changes can only apply to the drivers accepting them, not to all. There is at least another country where the court decided that any changes apply only to the plaintiff. The case of each driver must be looked at individually.

Moreover, it could be argued that the Supreme Court dismissed important factors in its decision, such as the potential repercussions of the ruling on Uber and its business model. By requiring Uber to treat drivers as workers, the company will be forced to pay additional benefits and protections, which may result in higher fares for passengers, more than 25% withholding (in excess of over 50% of the fare sometimes), and way less pay per mile for drivers. This, in turn, could negatively impact the livelihoods of drivers, who rely on the income from fares to make a living.

There are clear indirect allegations that a higher power from the government may have exerted pressure on the Supreme Court to reach its decision. This raises questions about the independence of the court, and whether or not the ruling was based on the best interests of all parties involved.

In conclusion, the UK Supreme Court's decision to treat Uber drivers as workers has sparked a heated debate about the status of independent contractors in the gig economy. While groups of interests see this as a victory for workers' rights, others believe that the ruling ignores important factors and the will of individual drivers. The damage caused by this ruling has damaged the life of millions across the UK, both riders and drivers, including the overall economy. No one is willing to voice these concerns, while the situation is aggravating from one day to another. Drivers do not want to get licensed and become partners with Uber, existing drivers are quitting en masse, riders are struggling to get from A to B where there are no other means of transportation available, and ultimately lives are being put in danger, as affordable transport becomes a thing of the past. The Supreme Court disregarded the most important implications when upholding an unlawful ruling of the High Court.

13. The Good Old Taxi Industry

It is a well known fact that taxis did not know how to move on with the times. Otherwise you would not still see taxis that are 20 years old picking up passengers for 3-5 time the price of an Uber ride, meaning that you will most probably get picked up in a very old vehicle. Of course there are hiccups with

Uber too, but the experiences cannot be compared. And when it comes to black cabs, they are just a bus ride on a smaller scale. Even the new ones (electric and hybrid) are equipped with stiff seats, and the good old yellow bar to hold on to when being thrown from one side to the other.

We like tradition, and London should not lose it, but caring for everybody having access to affordable and comfortable private rides, any place and anytime, is more important than not disrupting the "luxurious" black cab industry of the metropolis bourgeoisie. And yet, while Uber drivers have had their incomes reduced by 50%-60% overnight by the Supreme Court's ruling, black cabs' income went back to £3,000-£5,000 a week, as confirmed by retired black cab drivers. Yes, if you didn't know, even active black cab drivers are booking Ubers, because they are cheaper and more comfortable. You should listen to retired cabbies about how they were driving tourists around the block to increase the fare. They know the way alright…

During our first year as private hire drivers, we had lots of trouble with black cab drivers in central London. They all calmed down at the beginning of 2019, after some of them lost their licences for abusing drivers. As we said it before, we do not want London to lose its tradition. That is not possible anyway, as cabbies have their own clientele, from politicians, superstars, etc. These are also categories of riders that tip no matter the cost of the trip. One of us had a brief relationship with a British lady who was only booking black cabs.

Just try to understand the overnight difference between the earrings of Uber drivers and black cab drivers. The first are struggling nowadays to reach £900 per week with UberX, Uber Comfort, Uber Green, and Uber Pet (usually driving a Tesla model 3), no matter the hours put in, while a black cab is earning in excess of £3,000-£5,000 per week in probably the same amount of hours. They also received non-refundable grants of £50,000-£60,000 from the government to buy new vehicles, during the pandemic. In return, most of them stayed home throughout 2020, when Uber was people's only option in most areas of London. When knowing all this, who do you think was more entitled to the non-refundable grant?

What people need to understand is that there is enough money in the market to go around for everybody. This is how the economy works, someone with a better idea or business model comes in and steals from the other one that fails at evolving. What should all the pizza places in London do? Kill each other until only one pizza place is left standing? There is enough money for everybody! Over regulating private hire with political undertones, just to protect the black cabs, or local cab firms (this one is the most outrageous), and in such a gross and evident way, is proof that we have probably never lived in a democratic society. If you cannot count on the highest court in the UK to uphold democratic values, there is no way you can expect it from the

Parliament, as they are the cause of all intrusions into private life and affairs.

Our question for taxi drivers in general is, where were you during 2020? Where there were taxis galore, usually at train stations, all taxi ranks were empty. We could only see other Uber drivers picking up and dropping off, while patients (ill with the new coronavirus by the way), doctors and nurses were desperate to attend patients, and telling us that Uber was their only available transport option. Uber even gave medical staff free rides at the peak of the pandemic, while reducing the drivers' pay per mile, and the only ones taking the high risk were us, the drivers. What happened to the most greedy taxi drivers during the height of the pandemic, and why isn't this being debated? We do not consider ourselves heroes, but in the context of us risking our lives and our families' lives, did we deserve such mistreatment from the Supreme Court, unions, Uber and others, politicians, taxi drivers, and the people we serve? Did we deserve not being listened to individually, while the scaredy cats of the pandemic received brand new vehicles (courtesy of Sadiq Kahn's lobbying) and have had their incomes only going up throughout these years? It is the riders' decision who they want to ride with, and if some have turned away from taxis, it is their own doing. Every human is free, and entitled to perform in any industry, no matter what the veterans think, and Uber has been highly beneficial for the economy compared to the taxi industry. Why? Because more than 75% of the money on the platform used to go back into the real economy, which is supporting it in a healthier way compared to paying a lot of corporate and income tax. And with Uber there is no cash involved compared to taxis, where cash is still preferred.

Any Uber driver knows that their downfall has been supported and accelerated by the taxi industry, via their political connections in the government and the Mayor of London, who was always bad-mouthing private hire and Uber drivers in particular. The fact that you can "sponsor" MPs to support you in your cause is conflict of interest and bribery, which should involve a criminal record for both parties, and the dissolution of at least one large union. Directly causing and aiding the financial collapse of tens of thousands of drivers across the UK is another crime in itself and a shame for any government that "wants to save" the economy. The state of the economy is due to most people living on less, not on more. Reverse it and you instantly have a robust, working economy.

14. Uber During The Pandemic

It's a fact that Uber drivers' financial downfall happened during the worst ever pandemic and it is astonishing how coincidental it is with governments around the world taking measures to reduce people's incomes (non-pandemic related). And whoever does not have political connections, falls. In this case Uber drivers. We are not a guild, and we do not care for other drivers in most part. An independent artist on Fiverr does not care about another independent artist on Freelancer. It is the same for taxi drivers, but since most of them have decided to do it their entire lives, they're a tad more consolidated as a group, but only if private hire is around, meaning some sort of competition. You don't really think that the "Fake Taxi" guy cares one bit about the image of black cab drivers... Our common goal is to make more money than the other driver, whether they're another private hire driver or a taxi. The same is valid for taxi drivers, as confirmed by them. They may be seen as a guild, but they are not in the true sense of the word. Because they have been around for a while, people might associate them with the image of a united and consolidated group, but that is true only if competition is around. The fact that together contribute to pay MPs for legal support does not make them a guild. We do not think that as self-employed you can form a true guild. There is no guild of CEOs.

This is what the Supreme Court has done to Uber drivers – it considered us a guild, therefore employed. Here lies the contentious of the ruling.

During the pandemic, Uber offered free rides to medical staff at the expense of drivers too. As a huge high income corporation, they could not afford to do it at their expense alone. We're saying high income because even at 25% service fee charged from its partner drivers they were making huge money, as they have little staff in each country and producing no physical products. With all this, Uber did not bother to create a PR campaign meant to showcase the bravery and sacrifice of their drivers, at any point of the pandemic or after. We are some of the highest rated drivers in London and what we've always seen in the media or on TV about us was negative. How is that even possible, when millions of trips are being finished in London alone each month? Even unhealthy food companies like McDonald's are pumping billions into convincing consumers that their food is the best. But not Uber, when the whole company was in the first lines during the pandemic through its partner drivers.

Why did we continue to work during the pandemic? The first reason was because we were not afraid of any virus, and we wanted to help as much as we could. Throughout the pandemic we did not hear of Uber drivers dying from this virus, despite being the most exposed group of people out there, therefore we were right to not be afraid. Since no taxis were to be found on the roads, we knew that Uber was now paramount to the success of securing medical help in hospitals. Doctors and nurses didn't even have public transport available in most parts of London. This says a lot about the unspoken truth of how the government cared about helping ill people, no matter the reason for their illness. A lot of the doctors and nurses were calling us heroes. A second reason was because we could not afford to stop working, as this is our main source of income, despite owning separate businesses.

Our ability to pour money from driving with Uber into our side ventures had been cut short in March 2020. The business could not even ensure our survival, and negative personal credit grew bigger and bigger. We didn't even qualify for the government grants due to the rules for qualifying. So, we had no other choice but to struggle, and make things happen with barely £50 of income per day, and sometimes even £8 in 12 hours of online time on the platform. A third reason was due to wanting to support Uber in going over this crisis, making sure that the service does not disappear by the end of it. We knew that it will eventually go away (little did we know that it wouldn't), and that rider demand will increase with at least 30% compared to pre-pandemic levels. And we were right again. The surge happened, but we got pulled back by the implementation of the Supreme Court's ruling (to treat us as workers), and by Uber's audacity to keep more from the fares, instead of paying the benefits from the 25% they used to charge before the changes. This contributed to the nightmare the drivers are living in nowadays, despite the massive surge in demand, earning somewhere close to the pandemic levels of 2020. Welcome to the new pandemic for drivers and riders – overregulation and government control over corporate operations. Don't be fooled though and adopt a skeptical attitude toward what we're describing here. The government has done it to other self-employed people during the pandemic as well. Depending on the sector, they've forced them to work through third party companies to get guaranteed benefits that the regulator sees fit, and the independent contractor is earning less money because of that. Basically the government wants independent people to get benefits, despite this undesired intervention hurting their overall income. No matter what you say, the plan to manage your personal life without consent goes on. The pandemic was about more that just a virus going around…

It was not all for nothing though, as Julian managed to write "Kiss the Girl Save the World Kill the Baddie", a non-fiction novel about his life and what happened during the pandemic from a ground view perspective. In 2022 he also published "Coming & Going", his first volume of poetry. You must not let a crisis go away unexploited! That's the only good thing we've learnt from the government so far.

15. S.O.S. (The Economy)

We need to understand that the two most unhelpful groups of people for the economy are the ones living on benefits and the employed. You're probably raging while reading this, but hear us out. The Economy is always in a bad shape because people do not have financial stability. If you'd have financial stability, you wouldn't need a job, you'd be at least self-employed, working one contract at a time and earning a lot per job. The vast majority of employed people are living on the edge of survival in most fields of activity. But they are paying tax, which goes to sponsoring wars, fake charitable acts, helping other countries, keeping many companies alive with recurring poor repairs, supporting a medical service that can never function properly, bailing out banks, pharmaceutical and energy corporations, etc. If you believe that paying tax is a good thing, you've got another thing coming at you. No! Paying tax has been proven over and over that it is not a good thing, as corruption will ever be present in a centralised authority (government). Otherwise, everything should be good, and many issues temporary, with huge time gaps before them recurring. The unique solution is having more decentralised/independent individuals (most of society, if not all), using their skill to help others, while for eventual conflicts a spontaneous council should form and dissolve after finding resolve. We are living in and building a sick world full of too many employed people living on the edge, with the number of very poor increasing rapidly. We are pretty sure that you, the one reading this, have a job but are still struggling, and if you stop working even for one month, you'd be in big financial trouble, more or less. So you're willing to compromise. With what? With your freedom. You'll do as asked by any big corporation or group of corporations (monopolies) and the government. If you don't, you get buried even deeper. As self-employed, you don't care about any corporation or government. You're your own boss, earning big per each contract, going travelling, living life so to speak, and paying your share of tax. In opposition, by not being able to catch up with inflation as employee, in like ever, due to the US

Dollar coming off the gold standard in 1971, you won't be able to concentrate and work towards your full decentralisation; you'd be focused only on how to make more money, how to survive and fight off constant decay. Because the final goal for every human should be to not needing to use any form of money altogether. How many of you have this planned for your future? We would say none, because you were programmed to think that the world cannot function without money. Whereas money is just a means of control and for now the only way for you to achieve complete freedom.

As self-employed, in 2019 Uber drivers used to make £6,000 per month, compared to 3000-3500 or even less today. As independent contractor, you're almost always way ahead of inflation, which is a state of affairs the government does not want. This is what they will do to you if you accept using a CBDC. They'll get you poor and non compliant, and then they will restrict your life fully. You will only be allowed to just exist. More independent contractors means more business diversity, while more employed people causes less diversity, an ossification of creativity in society. Everyone will have to create around big corporate ideas/ideals only. Look at how corporations on the likes of Google and Microsoft have started and what they have become. And now Uber joined the same tyrannical trend, pushed by the government and justice courts, while its executives are being named from the ranks of conspirators, not chosen based on skill.

The gig economy has been growing rapidly over the years, maybe too fast for some people's likings, with companies like Uber playing the biggest role in this trend. Uber is a platform that connects riders with drivers, making it easy and convenient for people to get around. Despite the controversies and challenges that Uber has faced, it is undeniable that the company has had a significant impact on the economy, for all the reasons above, and more.

Firstly, when drivers earn money through Uber, most of that income (75%+) was going back into the real economy. This is because drivers are independent contractors, rather than employees, and they are responsible for paying for their own expenses, such as fuel and car maintenance. This means that the money that drivers earn through Uber is not just being absorbed by the company, but is being spent actively in the local economy. This, in return, helps to stimulate local businesses, and create jobs. Furthermore, when drivers are able to earn a steady income through Uber, they are more likely to invest in themselves and their businesses, further fueling economic growth.

Secondly, while it is true that Uber was paying tax on less than 25% of its entire income, it is important to note that the company was following the tax laws that were in place at the time. The tax laws for the gig economy are still evolving, and it is possible that the tax obligations of companies like Uber did not require them to pay tax on most of that income, due to company expenses, etc. Regardless, it is important to remember that Uber is still paying taxes, which is more than can be said for many other companies that engage in tax avoidance, like oil companies not paying tax on billions of pounds. With Uber, most of these billions go back into the real economy.

Thirdly, if Uber were to disappear from the economy, it would have a significant impact. Not only would this mean that hundreds of thousands of drivers would be out of work, but it would also mean that millions of riders would be without a convenient and affordable way to get around. This could have a ripple effect throughout the economy, as businesses would suffer and people would be forced to find other transportation options that may not be available most times or really expensive to use. In this way, Uber has become a critical component of the modern economy, and its disappearance would be felt by people and businesses alike.

Finally, independent drivers are better for the economy than employees or people on benefits. When drivers are considered workers, they are typically paid a set wage, regardless of how many hours they work or how many rides they complete (current state of drivers). This means that the company should be taking on all the risk and cost associated with paying them, but drivers still have to pay for everything else to be able to work, which is actually against the worker's law. On the other hand, when drivers are independent contractors, they are able to control their own hours and earnings. This means that they are more likely to be motivated to work harder and make more money, which is better for the economy as a whole. Additionally, when drivers are independent contractors, they are not eligible for benefits, such as health insurance or paid time off. But this is not an issues if drivers earn a lot to cover for it themselves. This means that the cost of these benefits is not being absorbed by the company, which allows it to be more competitive in the market.

While the gig economy is not perfect, it is clear that companies like Uber are making a positive impact on the economy and the lives of people all over the world. Why would you try to destroy it?

16. The Silent Voices – Riders

Here we are today, in 2023, not being able to book a driver on the Uber platform, or any other ride sharing platform. It can take hours to get a driver to take you to central, south, or south-east London. It can vary, and might actually take hours to book in other situations too. You might think that there are no drivers around, as per the news in the media, but suddenly you see a wave of vehicles with the yellow circle in the windshield and on the back door window driving past you, or waiting parked on the side of the road. "What is going on?!" After an hour, a driver eventually accepts your trip request, but he cancels. "Damn it! What the fuck!!!" You go to the taxi ranks and ask them how much they would charge you to go to Dunstable and they say £90 for 30 miles. Uber was quoting you £40. You laugh and go away. Everything is frustrating and you are so late you feel like crying. What is there to do? Finally, a driver comes to pick you up. You start chatting and find out that from that £40-£45 that you're paying, the driver only gets £20-£25. You think that its not that bad, as that's a 35-45 minutes drive. Well, it is really bad for the driver because Uber is screwing him over by withholding more than they should. He doesn't want those imposed benefits, he just wants to drive as before, with Uber keeping 25% or less. And for an Uber driver to drive with less than £2 per mile is really bad, considering how all the costs have skyrocketed. Transportation should always be good money! Uber is a private service, many times UberX sending vehicles that look more like Uber Executive, with panoramic roof, your desired climate and music, really new, clean, a very safe driver (with many thousands of trips under the belt), while no cash is involved. If you don't appreciate this service and want to go back to paying £90-£100 for a 30 mile trip, in a 20 years old taxi with a driver smelling of cheap tobacco, then go ahead and do nothing about bringing back the glory of pre-pandemic Uber. Don't support private hire drivers. Of course there are bad apples on the Uber platform too, but they are a very small minority. If you want to be able to get a driver instantly when you need a ride, you must stand tall as rider, and demand action from the government to fix it. Because you are in charge when it comes to which service you prefer to use.

There is so much you can do as a rider, but the small things you can do can mean a total redress of the situation and a new golden era for drivers and private hire in general. Uber and other platforms should not withhold more than 25% of the fares, and never treat drivers as workers, as this is against existing laws passed by the UK Parliament.

There are a few ways of action that riders could take. Firstly, every corporation the size of Uber needs to justify its actions in front of its users. They need to be listened. You cannot get someone dependent on your services and then crash your business. For at least ten years, people have adapted their lives around private hire services, i.e. Uber. We even had riders that sold their home and moved to our area in order to benefit from Uber services. They were shocked when everything changed and saw that it takes sometimes more than an hour to book an Uber for London. Changing areas to be able to use Uber is a huge commitment. This is why, as user of the Uber platform, you need to contact Uber and ask that the new system be reversed. You must ask that drivers are considered independent contractors, and the contract to be between you and the driver taking you from A to B. You must contact Uber directly, by sending individual letters, starting a petition and asking the Parliament to intervene, and even asking to speak with Uber's executives. You can do the same with Uber BV, which is the European parent company with its head office in Holland. In the UK, the company is called Uber London Ltd, with head office in Aldgate

Tower in Aldgate. There are millions of riders using the platform in London alone, so if all would take the above actions, things would change in record time. All you have to ask is that Uber defaults back to the old system to allow drivers to thrive and make as much money as they want, and not as much as Uber wants, which is barely survival. Remember that if this continues you will lose private hire and Uber as a service. All that will be left is the good old expensive and unreliable taxi. This means that society will go backwards instead of evolving into a thriving ecosystem for all.

A second complementary action is to contact and demand that Transport for London give up their overregulation pressures on private hire (Uber and others) and allow these companies to default to the old system, and focus on addressing the real issues that needed fixing. Write emails and letters explaining what is happening with your life because of how drivers are being treated. Do this into millions and demand that your freedom of choice is listened to. You want this service to work again for everyone, and that your drivers coming to pick you up are happy. Many have families and they jumped from having a great life to barely survival, while rider demand has increased considerably. Plus, there will be no switch to "green" vehicles if running costs are more than double of a hybrid-petrol. TfL must listen to you! Disturb them as much as you can until they take relevant constructive action to give people their freedom back.

If you live in an area with Uber service, contact your MP and demand action to redress the situation. Bear in mind that these people and institution will always try to make it worse for riders and drivers before making it right. You need to ask for the right things, otherwise you can make it really bad for drivers. Most MPs are freemasons so their solutions are not meant to make it right, despite pretending to listen to people. Kind of like what is happening in Oxford with the 15 minutes boroughs. "No matter what the residents are saying, the plan will go ahead." Take that! If you decide to demand the fixing of the Uber service back to what it used to be pre-pandemic, you need to pay attention to how you word you request and what you are actually requesting. Be smart, respectful, informed, and helpful.

Make sure you read the decision of the Employment Tribunal of the Supreme Court carefully, and then contact the court to ask for the cancellation of the decision, or at least its amendments to apply only to the Plaintiff (Yaseen Aslam, James Farrar, Robert Dawson, and others). The first two have stopped working with Uber before the filing of the lawsuit, in 2016, and never came back working with any of these ride sharing platforms. The "others" are meant to be all drivers that are part of a union supporting the cause of being treated as workers. We know as a fact that they are a minority of drivers (paid, misinformed, brainwashed, etc.), but to this day these unions have refused to disclose their number despite being contacted by us. They behave as if it's a state secret. We have a hunch that they are less than 1% of the total number of licensed private hire drivers. That means that a very small minority have ruined the lives of millions across the UK. Millions of letters should be enough to persuade The Supreme Court to amend the decision. Be very specific with how this has altered your life, as the highest court in the UK has a moral responsibility to do good, to improve human life and condition with its interpretation of existing laws. If any of their decisions have caused harm in any way, they want to know about it. The Supreme Court is not a politically correct institution, not officially at least. At the end of the book you will find all the details about how to formulate your letter and where to send it. In all instances send individual letters, not as a group. Group action is weak action! We know, this sounds messed up, but it isn't. Trust us, we've been there more than you and multiple times, and if it were for group action (union) we would've been in big trouble. You can read more about it in "Kiss the Girl Save the World Kill the Baddie". The letter presented at the end is meant for drivers, but you can adapt it to your condition.

You can also contact your local Council. Its mission is to listen to and serve you, to improve your life. Explain everything in detail, how you need and depend on Uber, and demand support from the government and TfL. The Council works directly with the local MP.

Contact all of the above simultaneously and make a stand. There is no need to protest physically in any way. This is silent action and if millions bombard them with correspondence asking for clear action, they will have to give in. It is incorrect and wrong to name them *authorities* because they are actually **delegates**. What you call authorities are actually people delegated by you and us to fix the issues in society, not to create them, and then roll back their power until new issues arise. You have the power. We took you from A to B in absolute safety and comfort. Where will YOU, the rider, take us? Fight back the smart way!

17. The Silent Voices – Drivers

Drivers are the backbone of the Uber business. Without them willing to become licensed and drive on the platform, it would be impossible for them to make any money. A driver's commitment is huge, as they need to spend about £600-£800 just to get licensed. Then they need to get their own vehicle, which needs to meet the standards. This implies that it shouldn't be older than ten years at first registration. There are a lot of rental companies partnered with these ride sharing platforms, but after renting from them for a while, we realised that it was not viable. This is because you need to make the money to pay the rent first, then to start planning for other expenses, and then think of profit. Almost all rental vehicles cost about £1,000-£1,500 per month. It didn't make any sense for us, even with the pre-pandemic level of earnings. If you decide to go on holiday for two weeks, your account will drain with lightning speed, as you have to pay the vehicle rent every week. Drivers that tell you that it is *ok* don't really put costs on paper and usually sleep in the car, while they go home just to VISIT their families. If that's a good and healthy life, then we don't know what life is. Freemasons like to say that life is a struggle, while they're enjoying a lavish worry-free lifestyle…

The above example is meant to emphasise on the commitment of each driver, just to work with a "trustworthy" company like Uber. If the standard in the market cannot be trusted, there is no need to reference the competition. So, it is not easy to get new drivers, nor keeping the old ones on the platform willing to do it anymore, if they are just barely surviving. This industry cannot work with Uber dictating us how much we can earn per month. To give you an example, we did not have a holiday since July 2020, compared to 2018 when we went on eight holidays. That was the last time we could afford one. Since then it was constant grind for us.

We have to understand that it is not brain surgery to drive people from A to B. Almost all drivers could do it. But it is not easy either. When you get bad riders, which can happen more often than you think, you feel like quitting, and there is no one there to compensate you for the stress. Yes, when you behave bad as a rider, we would like you to pay for the mental stress and for putting our lives in danger.

This is one important aspect that Uber had to improve – better monetary compensation from bad riders. We had riders that wanted to fight us, or who've touched the controls of the vehicle while driving with speed. We believe that situations as this should imply a significant financial burden for the rider. On top of that, how do you think an UberX driver feels when picking up someone that has a Ferrari or Lamborghini parked in front of the house, and takes them from A to B with less than £1 per mile, having to meet Uber Executive standards, and they don't even tip at the end?

As a driver you go through a lot. This is why you have to fight. You are an independent contractor, and you can never be a worker, nor should you ask for such nonsense. Uber would seize to exist (which is not a far-fetched allegation). It's like scoring in your won goal. Drivers come from almost all spectrums of life, some are more educated than others, and others wiser than the rest. What unites us (not in a union but in cause) is our drive to be independent, to work when we please and earn as much as it is available in the market. Why would you search for financial independence, and then ask to be treated as an employee? Does this make sense to a healthy mind? We truly believe that all drivers that supported these changes to our status on the platform should be banned from carrying people, or at least be psychologically assessed by the toughest professionals. If you are reading this book we take it you're not in this category, but if you are, you need to turn inward and check who you really are, and why you believe what you believe. Because there are ways to turn people to the dark side. One is the good old Cuban-cigar-smelling money, and the second is the psychological weapon. A third is by using frequencies to alter your brainwaves, but that might be outside of the scope of this book. Which one is you?

If you are a driver who wants things to go back to what they were, follow all the steps recommended to riders. Take the same cumulative actions. Do it individually, not as a group! You have all the resources in this book. Use the letter at the end, shape it around your case (don't send it as it is), and send your complaints to the Supreme Court at the address provided. The court does not know that you are suffering, but they want to know if you do. They want to fix it, but if no one complains the smart way, if no one is contacting them, then it means all is good. By not taking legal action it means that you are an accomplice to the whole situation. Don't be complacent. Be proactive. If you decide to do nothing, then you deserve what's coming toward you.

All drivers must demand a full audit from Uber and any other platform they work with. Ask it by contacting Uber's management directly at their main office in London, or through a small court. Even before the pandemic there were systemic inconsistencies when it came to being paid the full tip, not getting paid after the wait time, Uber's 25% share, how much they were charging riders and what you were being shown in the app, etc. Drivers never saw the receipts issued to the riders at the end of each trip. With the new system, these inconsistencies have turned into gross misconduct and total disregard toward transparency and fairness. Many times, drivers are being shown something and then paid something else, they cannot see how much Uber is charging the riders, sometimes they pay for traffic delays, other times they don't, they pay you less when taking a shorter route, and so on. It's a Wild West out there at the expense of drivers. A full audit of all accounts will show a tremendous mishandle of money on the app from Uber and the competition. During the audit, Uber needs to show the full GPS data of each trip, all receipts sent to riders after the trip plus tip receipts (they are separate if tip given later on), and all other data related to the trip. Uber is a Big Data corporation, meaning that they have all this data recorded. This is what they're actually after – YOUR DATA! IF they get there in one piece, they'll use it in the fully autonomous vehicles. So, drivers should be compensated massively by companies like Uber for contributing to this technological advancement. This is why it is very dangerous that they've been taken over by our **delegates**, the government, through overregulation and by naming Uber's executives. Ask for an audit and get your money back. You will also have a case for financial compensation with the High Court.

We are aware that not all drivers are silent in these matters, but just complaining and not taking action means nothing. It'll just turn you into a sour driver that no one wants to ride with. Stop spreading this negativity around you, and do something. You've got all the advice and examples you need in this book alone. And please stop throwing yourself into the bear's claws by taking group action, or even worse, joining one of these unions. They are using you and are run by people you do not want to have around. Plus, they will negotiate with Uber behind your back and for all the wrong reasons, and you'll end up really poor and bitter. If you've joined Uber before the pandemic, you know how it used to be. If you've started during 2020 or later, what we can say is that if you were working with £15-£20 per hour (online time), that was a bad day. We're exhausted and we need a holiday really bad. Plus, we can no longer afford to put money into our side gigs, nor time to work on them. Let's bring the good old days back!

18. The Silent Voices – MPs

As the use of ride-hailing services has become increasingly prevalent, politicians and Members of Parliament (MPs) have also started to use these services. We even had senators form the US booking us to Westminster. This means that MPs are familiar with the various issues that Uber drivers are facing and are in a position to take action to address these problems. Most probably it is highly difficult for them to book an Uber too. Whether they see the Supreme Court's decision to treat us as workers a good thing or not, they should not overlook the damage it has caused. MPs are well aware of the reasons behind the legal action that has been initiated against Uber, and they should come forward to explain it to everybody. The reasons behind the cancellation of Uber's licence the first time were outrageous and a coverup for TfL's failings. The fact that they are not publicly discussing the urgent need for addressing the issues plaguing private hire after the drivers started being treated as workers, and that these changes were demanded by a handful of drivers, is in itself a covert operation against drivers to which MPs are definitely a part. This issue is too big to not get noticed by them. If you are an MP, come forward and support the recovery of this industry. It is not only about the struggling drivers, but about the millions of riders depending on the service.

19. The Corrupt Unions

Who can start a union? Well, anybody. But in reality, a person with certain connections will start a union. Same as charities, they are businesses in the true sense of the word, even though they're not limited companies. They live from the membership fees, but in order to take legal action against a corporation like Uber, a lot of money needs to flow in from other sources. We won't discuss about the GMB union here, as they are too big to not be corrupt. We'll debate ADCU, the union started by Yaseen Aslam, James Farrar and Robert Dawson, the initiators of the lawsuit for worker status. Yaseen Aslam was complaining that he could not support his family while working for Uber before 2016, but after the start of the legal proceedings he somehow had enough money to survive, support his family, and pay for top notch legal representation. Hmmm... At least interesting and suspicious, isn't it? At the time ADCU was a limited company with a handful of members. We have our own experience with unions being run by undercover agents, but in this case we'll give them the benefit of the doubt. At the end, even Forbes wrote an article portraying Yaseen as a hero. All while other drivers of the same culture as him didn't want to shake his hand when meeting him in London. It came to our ears that some spat at him for what he's done. His union shows as dissolved on Companies House, and probably that is the moment when they got registered as a union. They are still very much active, and protesting against Uber, in a room, and sometimes on the street, as shown in online videos. We're guessing that's his side gig when taking a break from his IT contract with the MOD (Ministry of Defence). We do not know too much about James Farrar, but we're inclined to think that he was never a driver with Uber, as we've never seen a white British man driving on the platform in a Prius (not that anything would be wrong with that). He looks more like a black cab driver dealing with the "problem" that Uber was for that industry himself. The most undercover of all seems to be Robert Dawson. ADCU has an online presence and if you feel like investigating yourself, just have a look at how many followers they have, the people supporting them, and what they're fighting for. We've done some more research on them to discover that they had only about 4,600 members at the end of the fiscal year 2021. This means that a minority of drivers changed the business for everybody, since there are roughly 50,000 drivers or more on the Uber platform in the UK. They are not replying to emails or the phone number provided either. It is true that the situation is clearer for a trained eye, but with a bit of effort even a novice can see this union as it truly is.

We like to think that no driver is supporting these conspirators because they've caused enough harm, but that is probably not true. We think now that most of their members are drivers that started driving during the pandemic or after the changes to our status was implemented. Today this union may have more credibility in the eyes of these drivers, as they don't know that ADCU was the problem from the beginning.

Finally, the fact that a union was established by two or three individuals just to attract attention, they won the lawsuit against Uber using funds from questionable sources while its founders were complaining about not earning enough money to survive, and then they destroyed the business for tens of thousands of drivers without asking for permission, all of it calls for a criminal investigation. The destruction they have caused is similar to terrorism, as they hurt millions and did not contact all drivers to get their opinion. Even more aggravating is that they painted a picture as trying to represent all private hire drivers and app couriers.

The main take for the reader should be to never trust unions. A union is the government's way to control the masses of employees. There are plenty of ways to take legal action yourself, you just have to educate yourself on the matter and do it. A union will either ask you to compromise with the company or the government, although you haven't done anything wrong, or get you a mere wage increase that is always under the level of inflation, or neither. If you are part of a union just make it your mission to really find out who the leaders are in public and private life. We guarantee that you will be shocked by the findings. Many of them are members of secret societies. If you have a bit of money you can hire a private detective. Some of them have connections with the police and other services, as they might be former police officers or other types of agents. They're not cheap to hire but you can join with others to make it easier. The information you'll receive about the union leaders will be very juicy. Just be careful so they don't find out. Make sure you really know the people you're joining forces with. In the past we made this type of mistakes, and they didn't hold back from hurting us. If your private detective disappears, we suggest you pull out of the union and move on.

Do you think it's fair that 4,000-5,000 drivers, union members, have destabilised the entire private hire industry (50,000-70,000 drivers) on the Uber platform, the standard in ride hailing services? Why do you have to suffer for the minority's views on business, freedom, and quality of life?

You must understand that there must not be one leader for all Uber drivers. Any Uber driver must be his/her own leader. This is what you've signed up for when deciding to become self-employed. Regulators want a leader for all because they can easily be manipulated and corrupted. If you're a leader for your cause they cannot control an unlimited number of leaders. This is why unions "fail" in obtaining benefits that will be grater than inflation.

20. Workers

Workers law is a set of regulations and policies that are designed to protect the rights and interests of employees in the workplace. This law is often seen as a necessary tool for ensuring that employees are treated fairly and that their basic needs and rights are respected. However, despite its importance, many people believe that worker law does not provide adequate protection for the self-employed. These are the people that cause overregulation and ruin lives. Self-employed people do not fall under the worker law, because they cannot ask for protections from their business partners. They work one contract at a time, earn as much as their abilities allow them, manage their finances with more skill, and can book as much time off as they want. Being employed or self-employed is a choice, and it comes with risks and rewards. Workers law is there to protect EMPLOYEES, and not independent professionals.

The moment the government intervenes in the lives of the latter, their activity gets damaged, which happened to Uber drivers. This mishandle of regulatory powers should be taught in universities.

It is wrongfully perceived that the self-employed often work without a safety net, without the security of a steady salary and without the benefits of health insurance or paid time off. This can be a daunting experience, and many individuals are reluctant to take risks as a result. Although this may be true to some degree, when an individual decides to become independent it means that they have acquired finance management skills that most employees do not possess or don't want to develop. In contrast, workers are often seen as being more risk-averse, preferring the fake stability and security over the potential rewards of entrepreneurship. There are different types of people in the business environment, and the ones choosing to be free should never have damaging safety nets imposed on them via regulators. If that happens, the only result can only be the destruction of the independent contractor's overall business. In the worst case scenario, the self-employed person should be consulted and their choice respected.

Workers are often thought of as individuals who do not have control over their own lives, who must follow the instructions of their employers and who do not have the freedom nor the ability to make their own decisions. In contrast, self-employed individuals are seen as being more self-reliant, taking charge of their own careers and making decisions that best suit their needs and goals. For the second group, overregulation of their activities is what they are afraid of.

Workers are also often thought of as individuals who wish to retire from life. That is not possible without dying first. They are seen as people who are ready to give up the hustle and bustle of work and who are looking for a comfortable and secure retirement. In contrast, self-employed individuals are seen as individuals who are always looking for new opportunities and for ways to grow and expand their careers.

Workers law is designed to protect employees, not self-employed individuals. This means that self-employed individuals are not eligible for the same benefits and protections that employees are entitled to receive. For example, self-employed individuals are not eligible for lack of work benefits, workers' compensation, or health insurance through their business partners. Otherwise, the entire business environment wold be destabilised, with dire consequences for the economy.

Finally, workers often think that a good life comes after 70. Usually, at that age you realise how wrong and misled you've been, and start contemplating how good it used to be when you were planning for retirement. How good can life be after 70, compared to the previous 50 years, regardless of how much money you have in the bank? The apogee of a holiday in Marbella is you eating in a restaurant and then having a walk on the promenade, while trying to stay away from energetic youngsters that might push you by mistake and damage your already frail hip. Workers/employees believe that they must work hard and save for decades in order to have a comfortable retirement. In contrast, self-employed individuals are often driven by a desire to create a good life for themselves NOW, not in the distant future. They are more focused on the present, and they believe that they can achieve their goals and aspirations through their own efforts and hard work. It's not that employees don't work hard, but they do it for other people, for the wrong reasons, and for a very little reward. The future can/will never belong to the workers/employed.

If you were part of the group asking to be treated as worker by Uber and other companies, first of all you're an idiot. Second, you should thank God every single day for keeping you a live, because that is truly a miracle, considering how little you think, how little you know about the world around you, about the world as a whole. If you want to be a slave of the rich forever, if you've got no "cohones" to dream big, to dream of being free one day, not needing money, nor to be ruled by any other human, do it alone. Don't drag others like us into the pit with you. You deserve what you're living now, what Uber and others are doing to you, how they turned you into their bitch, on your request. We don't deserve it because we never wanted these changes. There is a chance for your redemption, there is always this chance, but will you take it? Do you have courage to take legal action against the ones that lied and deceived you, Yaseen Aslam, James Farrar, and Robert Dawson, etc., and all other unions that supported the action? Despite your stupidity and what you've done to most self-employed drivers, we feel like thanking you in a way. While you'll struggle to make ends meet for a long time from now, we will make a lot of money with companies like Uber, especially due to what you've done to us. The only way to make up for all this damage is to take legal action against the conspirators, the ones you trusted to get you "worker benefits". You see, all has a resolve, and what goes around comes around. And this book is just a tiny grain of sand compared to how big this will get. And we will have nothing to do with that. We'll just be there to watch you struggle for survival. Say hello to karma for us!

21. Uber's Business Costs

If you've ever complained about companies like Uber not paying enough tax (on 25% or less before worker status), thinking that this is the reason as to why the economy is always in a bad shape, well, you are wrong and obviously don't know much about how businesses work, legally. That is an employee mentality right there. The economy is in a bad shape because of tax money mishandling, too many employees and benefit claims, and not enough independent professionals. The majority of employees will never be paid enough to live a worry-free life, and that is the main problem.

Uber's business costs are huge, despite having very little employees worldwide, and that is intentional. In London they have a Greenlight Hub for onboarding, and main offices in Aldgate Tower. We assume they have less than 200 employees in the UK, while milking billions from this market alone. They also don't produce anything, as they are a tech company. Their biggest expenditures are with licensing (we believe it to be a very inflated cost by TfL), probably huge donations to political parties and/or MPs, and the biggest of all, the programming of the two Uber apps (rider & driver), including servers, etc. Why didn't they pay a lot of tax? Simply because they could pay for all of the above costs as much as they want, and then pay ZERO pounds corporation tax. The biggest chunk of tax would only come from employees salaries, including executives, and HMRC can't do anything, as it is within the legal lines set by them. And these cannot be changed to charge Uber more corporation tax on what they don't owe. This would mean huge lawsuits and more money lost by the taxpayer than earned. So, people calling for more tax for Uber are either conspirators, agitators, uneducated individuals, people that want Uber gone, or just plain reckless. And why is this important if, for example, 75% or more (including Uber employee wages) from £100 billion per year goes into the real economy (which is far better than paying tax)? Imagine the benefits of £75-£80 billion pounds actively growing a healthy economy, compared to £1 million in tax or less going to HMRC. On the other side, the government will always want more tax to sponsor wars and build armies. Do you still want to force Uber to pay more tax? It's the same as asking to be considered a worker when you're an independent contractor. You'd kick yourself in the nuts driven by political agendas. This is why financial education is what every human should own. Jewish people are laughing at you for this, as they're teaching their kids accounting and finance from a very early age. And who controls the money flows in the entire world? We'll let you figure this one out.

With Uber now treating drivers as workers, even less money goes to HMRC, and way worse than this, less than 50% of it is going into the real economy from drivers. Uber's business costs can now be pumped even higher because they need to pay Adecco to manage the pensions (the contract is probably worth tens of millions of pounds per year). While Uber can withhold even more than 50% from fares at times, they're making even more money on the back of drivers, while disposing of more income in a closed circuit to avoid tax legally. This is what the regulators have done to the economy, and it is impossible that they didn't know what will happen in the light of these changes. They all colluded to do it, and somewhere on the way, a lot of money was promised and delivered to the right people with power. There can be no other answer as to why this has happened. All calculations are pointing to it.

So, if you're happy with drivers being now considered workers (and struggling for the first time in Uber's history in the UK), despite not being able to book an Uber anymore in many parts of London, no matter how much you're willing to pay for the ride, you are no different than those people in third world countries being happy that they're receiving medicines from pharmaceutical companies to clean their dirty water, instead of being helped with infrastructure to bring water to (maybe capture it from air) and clean it in their own house. Why give people more power, when they can have less, and be "happy"?

22. Uber's Rejected Trips

At the end of 2022 there were approximately 150 million trips being rejected by drivers in a 30 day period. This is based on the number of drivers advanced by Uber in London area alone. From an acceptance rate of no less than 95%, drivers are now rejecting more than 90% of trips. This translates into a 5%-10% acceptance rate. The delays in booking an Uber are now huge at many times. It is not due to driver shortage as the media is promoting. This unavoidable exodus of drivers is incoming still. The expansion of the ULEZ zone to the entire M25 motorway in August 2023 will force many drivers to admit that being an Uber driver is no longer a viable business under the current legal framework. Switching to electric is no longer feasible either.

Going into even more negative debt is financial suicide considering the market conditions and the uncertainty. The economical situation of the individual that has to work for money will get way worse before getting better. Forget about the overall economy. Remember that even when being told that the economy was booming, we were still suffering. We need to fix the economy of the individual!

But what about me getting a driver eventually, if I don't give up? The truth is that most riders will get a driver in the end but after huge delays, with your request being rejected by hundreds of drivers highly likely. Many drivers will eventually accept, being tired of waiting or panicking that they will make no money if they don't start accepting trips soon. One trip now, another later, and maybe Uber will send them a good trip/job. From a guaranteed opportunity, driving with Uber turned into a lame hope for making enough money to pay for your running costs, rent and utilities. When we said 150 million trips we meant the number of times trips are being rejected. Undoubtedly, some riders can never book a driver in certain areas, no matter how much they wait. When Uber is sending the trip request to drivers hundreds of times with the same price without paying more no matter how many times the trip gets rejected, drivers will turn red and reject it as many times they receive it, even if they don't make any money. Many trips actually save you more money and time than accepting them. And with this, Uber is now practically blocking riders from booking, even if they are willing to pay more for the ride. In other instances, riders do pay a lot and still can't get a driver because Uber is keeping too much from the fare, and is not willing to compromise with the driver. Uber doesn't want to lose a penny but drivers must. When there are train strikes, Uber drivers are making less money than during normal days. As you've probably figured it out, driving with Uber turned into a war of egos too.

We understand that what we're saying here is controversial, but it is based on facts known only if you're a driver, working for Uber, TfL, etc. They will probably try to sway you away from this picture by dismissing this data. If that ever happens, ask for proof. Also pay attention to their words, how they formulate ideas, and what they are trying to focus your attention on. Be aware if a clear answer was provided or not. Politicians and corporate executives have a way of talking a lot and saying nothing. Most times your questions still remain unanswered even after hours of debate and Q&A. This is the first sign of lying or at least hiding the truth. Be difficult to psychological manipulation.

23. The Consequences Of Uber's Imminent Downfall

Uber is so important for the economy that its crash will mean danger to human life. 2020 was clear proof of how important this service is to so many categories of people that had no other means of transportation available (taxis among them), including public transport. While even the government didn't care about patients, doctors and nurses, having easy transport available to go to hospitals across London, Uber drivers were performing as usual, helping more than they could even realise. "Sir, I think I might have covid…" "Hop in! Don't worry, I'm an Uber driver, I'm immune." This is what we kept on telling all our riders, and between March 2020 and November 2022 we had no illness, not even the usual cold, after being in close contact for long periods of times with tens of thousands of riders, some of which very ill and contagious not only with a coronavirus. Uber drivers have been the most exposed group of people in the whole of the UK. And none passed away from the new virus. The same drivers are still here driving. No wonder many doctors were calling us heroes. Who would've thought that Uber drivers could become heroes by default?

    If there where any isolated doubts over the impact on life and importance of independent private hire drivers working with platforms like Uber, they have been shattered in 2020, when Uber drivers were among the few that did not coward out of helping people at their worst. This is why they need to earn decent and sufficient money, not barely survival money. The fact that they were independent actually pushed them to help, for money too, but they could've just sit at home rewatching old movies on expensive streaming services, while being paid £700 a month in benefits. Without these drivers driving throughout, the crisis we all know would have been way deeper than it is, with many more excess deaths. And this is not biased or flawed computer modelling that never materialised, it is real world data. May God save us all in case Uber folds its business in the UK, and we're being hit by a similar crisis in the future. While new problems will surely arise sooner than later, the situation of Uber drivers can easily be reversed, and even supported with hefty compensation for now years of distress caused by all these factors.

24. Driver Costs Now

You need to understand that if people cannot book on the Uber platform, they will have to drive and therefore create more congestion, which is unproductive for the "green" agenda. This is how you know that they don't care about saving the planet. Making it difficult for Uber drivers to operate will result in a less cleaner air overall.

This is a comparison between a Tesla Model 3 standard range and a Toyota Corolla hybrid - 1.8 petrol engine. Another disturbing fact is that black cab hybrids are not being charged congestion charge, and most likely are less green than a hybrid Toyota (any model, black cabs are also heavier). This is a clear collusion against private hire.

<u>Tesla Model 3 standard range costs at £3,500 monthly income</u>

We'll take as example a rental vehicle because if you can afford a £50,000 vehicle, why would you want to drive with Uber and make a residual profit, if any?
Weekly rental: £275. **Monthly: £1,192.**

Battery size: 60 kWh. We'll use a median price per kW of about £0.50. As an electric vehicle driver you are being forced to charge cheap and really expensive. You'll not charge more than 60 kW/day but we'll use 50 kW per charge/day of work.
Price per day: £25-£30. **Cost per month: £750-£900.**

With electric vehicles, you need to consider weather conditions and that the overall mileage is being severely impacted by it. We'll add here a 15% variation, so the final monthly charging cost will go to **£862.5-£1035**. During cold weather you can even lose more than 30% of mileage in the battery.
Total median running costs so far: ~£2,146.
Rent & utilities in shared house: £900
Telephone: £22
Food: £300
Other: £100
Total: £3468
Profit: £32! If any.

As you might know, if living in London, these estimates are very generous. Your costs other than running your business can be way higher. So, in this situation you can barely survive alone when driving an electric vehicle on the Uber platform. If you get your Tesla vehicle from a dealership, the costs could be higher, as with rental companies taxi insurance and services are included in the weekly rent. On top of this, the monthly income of £3,500 is not guaranteed. We do not think there is a driver earning more on the Uber platform as of March 2023. With the old system, when drivers were considered independent contractors, this situation would not have been possible, while earning in excess of £6,000 in 30 days of driving. With the current model, you cannot afford to book time off or holidays. All benefits and tips count towards the monthly £3,500 cap.

Toyota Corolla costs at £3,500 monthly income
Weekly rental: £245. **Monthly: £1,062**
Fuel at £1.5 per litre: £550-£650 per month.
Total median running costs so far with a 5% variation: ~£1692.
Rent & utilities in shared house: £900
Telephone: £22
Food: £300
Other: £100
Total: £3,014
Profit: £486! Dangerous level of profit.
Profit difference between driving a Toyota hybrid and a Tesla Model 3: £454.

These are rough but generous estimates to favour profit, but they can vary greatly depending on each individual's living arrangements, servicing, tax, etc. With other electric vehicle models it can be even worse, as Tesla vehicles are the best performers among electric vehicles. When it comes to plug-in hybrids (probably the worst vehicles) the situation is terrible, as they are less "green" than full petrol vehicles. Drivers will eventually give up charging the battery, or charge it once in a blue moon, and drive it on petrol only. A plug-in hybrid is heavier than a full petrol vehicle (because of the battery), therefore burning more fuel for the same mileage. The fact that TfL is still licensing plug-in hybrids but not self-charging hybrids is a crime in itself, a damn right hypocrisy, a big lie, and anti-science. It is only meant to destabilise the private hire industry for the benefit of the taxi industry only. A self-charging hybrid is in electric mode over 50% of the time driven in central London, which is greener than electric vehicles.

These are costs that most Uber drivers do not put on paper before considering what has happened to the business and how damaging it is for them across the board. We hope this is enough to paint the overall picture of the situation of Uber drivers, in the context of more rider requests, higher prices for riders, a decreasing number of drivers (many are quitting), "benefits" for drivers, overregulation, 50%-60% overall income drop, switch to "green" vehicles, etc. It is a tragedy of gigantic proportions!

25. Driver Costs Then

We won't start all over again with the calculations. In order to understand, you just have to apply lower running costs (~30%, a lot has happened in the economy since 2019) to an overall income of about £5,000 with about eight days of time off per month. So, Uber drivers had about £2,500 to spare each month, a lot of time off to recover and go on holiday, and plenty of freedom.

All policies around the world are meant to reduce the population. No measure or change is being pushed forward without having this in mind, even if it looks as if politicians and regulators are trying to help. Once you develop the skill to spot the patterns and understand that many officials and executives are members of secret societies, you will be able to build your life in a different and more defensive way. You cannot plan for a good and healthy life by dismissing these factors. Now, Uber drivers are very poor, even poorer than Amazon workers (less take home profit), and that fits perfectly with the global agenda 2030 (poor, property-less, and "happy"). All policies deployed by governments worldwide MUST obey and meet the depopulation agenda set by a handful of people (council, committee, cartel), by criminals that want to take control of individual life. Dismissing this means that you are not capable of seeing the truth, or you've switched to a default state of blocking your cognitive abilities. You need to keep in mind that this book was written by two former border police officers that have seen a lot in their lives to validate the information provided. Many things left unsaid cannot be divulged due to safety concerns. You can check the information provided yourself by doing in-depth research about global policies and spheres of influence. There are thousands of official documents and books out there to inspect. Once you understand that behind the political left and right are the same people/entities, you are ready to plan your life the right way.

26. Driver Sacrifices

The emergence of ride-hailing apps like Uber has brought about a significant shift in the way people travel. These apps have become a convenient and affordable mode of transportation for many. However, the sacrifices made by Uber drivers to maintain the system are often overlooked. In 2020, when the world was grappling with the pandemic, Uber drivers risked their lives to help and save people, and yet Uber showed no gratitude for their efforts. Moreover, Uber drivers have also sacrificed their freedom and finances to improve the quality of rides.

Uber drivers have sacrificed their freedom by agreeing to work within the confines of the Uber system. They have to adhere to strict guidelines and follow the protocols set by Uber, which often limits their flexibility and autonomy. Despite this, they have continued to provide an essential and reliable service, ensuring that people can travel safely and conveniently.

During the pandemic, Uber drivers went above and beyond their duties to ensure that people could reach their destinations safely. They risked their lives to transport essential workers to their workplaces and ensure that people could access medical facilities. Unfortunately, Uber showed no gratitude for their efforts (as no hazard pay was paid whatsoever).

In addition to sacrificing their freedom and risking their lives, Uber drivers have also sacrificed their finances to improve the quality of rides. They often spend their own money to maintain their vehicles, and yet the company does not compensate them adequately. Drivers have also gone the extra mile to ensure that their passengers have a comfortable and safe ride. They have provided amenities like water bottles, phone chargers, and music to enhance the ride experience.

In conclusion, Uber drivers have made significant sacrifices to maintain the ride-hailing system. They have sacrificed their freedom, risked their lives, and risked even their finances to ensure that people can travel safely and comfortably.

However, Uber's failure to recognise and appreciate their efforts is disappointing and dishonourable. The company should show gratitude and compensate its drivers fairly for the sacrifices they make to maintain the service's quality and its continuity. If the current business model is kept as is, Uber will fail sooner than later. The current top executives at Uber in the UK and Europe are similar to Joseph Gentile, the Silicon Valley executive that crashes everything he touches, from Lehman Brothers to Silicon Valley Bank (SVB).

27. Rider Sacrifices

As Uber increases fares and makes it more time consuming to book rides, riders are sacrificing their money, time, and even their wellbeing.

One of the significant sacrifices made by Uber riders is the increasing fares. As Uber became more popular, the company has increased its prices to be able to keep more from the fares and make more profit under the umbrella of paying driver benefits. This has put a strain on riders' budgets, making it difficult for them to afford Uber's services. Moreover, Uber has implemented a surge pricing system that charges riders more during peak hours or when demand is high, while paying drivers less than decent per mile fares. This can further exacerbate the financial burden on riders.

Booking an Uber ride can also be a time-consuming and frustrating process. At times, it can take hours to book a ride, and in some areas, it may not even be possible to book a ride despite plenty of drivers being present. This can lead to riders being late for appointments or missing out on important events or meetings. In some cases, riders may have to risk their lives by walking in unsafe areas or using public transportation that is not as reliable as it is advertised.

Moreover, riders' wellbeing is also at stake due to the difficulties in booking Uber rides. Riders in need may be unable to access medical facilities or other essential services if they cannot book an Uber ride. This can have severe consequences, especially for those with health conditions or disabilities.

In conclusion, the sacrifices made by Uber riders should not be overlooked as well. The increasing fares, difficulty in booking rides, and risk to riders' wellbeing and lives are significant issues that need to be addressed. Uber needs to work towards providing fair, affordable, and reliable services that cater to the needs of all riders. This can be achieved by improving the booking system, paying drivers a minimum of 75% of fares, and ensuring that drivers are available in all areas. It is time for Uber to recognise and appreciate the sacrifices made by its riders and work towards providing a better service. As it stands today, Uber is going against its own idea of affordable and reliable transportation service for everyone.

28. Forced Switch To Electric

The push towards electric vehicles has gained momentum in recent years, with governments and businesses advocating for a more sustainable future. However, the transition to electric vehicles has not been without its challenges, particularly for Uber drivers. The mandate by Transport for London and Uber, forcing drivers to switch to electric vehicles, is placing a heavy financial burden on them. This has resulted in drivers making no profit, with no financial support in the form of discounts or non-refundable grants. Furthermore, electric vehicles have low mileage and do not have less time consuming charging capabilities (five minutes or less), making it challenging for Uber drivers to make ends meet.

The lack of financial support from Uber, the

government, or Transport for London has made it difficult for Uber drivers to switch to electric vehicles. The cost of electric vehicles is much higher than that of traditional petrol or hybrid vehicles, and there are no incentives to offset these costs. This means that drivers are forced to take on additional financial burdens, making it impossible for them to make a profit.

Moreover, the low mileage and lack of less time consuming charging capabilities of electric vehicles (with many drivers having their batteries software programmed by rental companies to charge to no more than 80% of their full capacity) make it difficult for Uber drivers to keep up with their schedules. They have to take more breaks to recharge their vehicles, resulting in lost income and less presence on the road. This is particularly challenging for Uber drivers who rely on a consistent income to support themselves and their families.

Additionally, the income of Uber drivers has dropped significantly due to worker status, exacerbating their financial struggles and hurdles. Many drivers have reported a 50-60 percent drop in income, making it more difficult for them to afford the switch to electric vehicles.

In conclusion, the push towards electric vehicles may be and look noble, but it is causing significant challenges for Uber drivers. The lack of financial support, low mileage, and slow charging capabilities of electric vehicles, are placing a heavy burden on drivers. Uber and Transport for London must provide more incentives, and unbiased real support to make the switch to electric vehicles more affordable and sustainable for all drivers. It is essential to consider the impact of such mandates on those who are most vulnerable in society, and work to find solutions that are equitable for everyone.

29. Communication With Uber

Effective communication is essential in any business relationship, and this is no exception in the case of Uber and its drivers. However, the communication between Uber and its drivers has been found to be lacking, with drivers reporting poor response times and a lack of constructive communication. Uber's use of preset scripts and disregard for drivers' demands has created a communication breakdown, leading to frustration and a lack of trust.

One of the major issues with communication between Uber and its drivers is the use of preset scripts. Drivers report that Uber's responses are often irrelevant to their questions and appear to be generated by a machine rather than a human being. This lack of personalisation and disregard for the driver's concerns, undermines trust and creates an impression that Uber is not interested in listening to its drivers.

Additionally, drivers who seek to correct payment errors have found that their accounts are flagged, which adds up to their frustration. This further reinforces the perception that Uber is more concerned with protecting its bottom line than ensuring that its drivers are treated fairly.

Uber is unapproachable when it comes to having to explain as to why they didn't listen to the majority of drivers that wanted the system to stay the same, when they were treated as independent contractors.

The breakdown of communication between Uber and its drivers is a significant issue that needs to be addressed. Uber must improve its response times and ensure that its replies are relevant and personalised to the driver's concerns. All in all, Uber needs to listen to the demands of its drivers and engage in constructive communication to build trust and foster a more positive partnership. Without meaningful communication, Uber risks alienating its drivers, leading to a loss of trust and ultimately, a loss of business.

30. What Needed To Change

It was imperative for Uber to improve communication with its drivers to foster a more positive relationship. This could've been achieved through personalised responses to driver concerns, and a more open dialogue between Uber and its partner drivers.

In addition to improving communication, Uber needed to provide more compensation to drivers in various situations. This includes compensation for the time spent waiting for riders, compensation for cancelled rides, damages, mess, and compensation for the use of personal vehicles. This will not only help drivers make a living but also incentivise them to continue working with Uber.

Furthermore, Uber needed to offer better protection to drivers against bad/dangerous riders. This could have been achieved by implementing an anti-alias system and providing drivers with the ability to refuse riders who did not have their real name in the app, as well as the number of their trips as it was implemented on Bolt platform. Additionally, Uber should've improved the fare structure to ensure that drivers are not being paid less but always more. This is possible by providing transparency in fare calculations and ensuring that fares paid to drivers are not subject to decreases.

31. What Needed No Change

Uber's implementation of the UK Supreme Court's decision, to classify its drivers as workers without their agreement, has sparked a huge and unnecessary controversy. Many drivers argue, for good reasons, that they are not workers and can never be because they have the flexibility to work whenever they choose and because they have the power over the contract. This move from Uber has led to a further divide between the company and its drivers, as they only considered the demands of a few drivers, disregarding the vast majority.

Additionally, drivers believe that they should

be paid at least 75% of the fares they earn to make a living wage. The fact that Uber has reduced the percentage of fares that are paid to drivers from the final fare paid by riders, has had a devastating effect on their livelihoods, leading many to question whether it's worth continuing to work on the platform.

While it is understandable that Uber is trying to comply with the court's decision, it is essential to consult with and meet its partner drivers who are driving people from A to B in the decision-making process. Drivers are the backbone of Uber's business, and their opinions should be considered. By ignoring the drivers' concerns and implementing the court decision without their agreement, Uber has alienated its most important asset.

Uber's decision to implement the court decision, without the drivers' agreement, has caused significant harm to its relationship with drivers. To avoid further damage, Uber needs to engage in open dialogue with its drivers and take their concerns into account. This includes ensuring that drivers receive a fair percentage of the fares they earn and recognising that they are independent contractors and not workers.

32. The Future For Drivers

Uber drivers are facing an uncertain future under the current conditions. Many drivers are not making a profit and are better off seeking employment elsewhere. The implementation of worker benefits has had a detrimental effect on drivers' income, and the worker status had the opposite effect of what was intended.

The worker status was intended to provide benefits such as sick pay, holiday pay, and the minimum wage for drivers. However, the reality is that many drivers have seen their income decrease as a result of these changes. The cost of providing worker benefits has been passed onto the drivers, leading to a reduction in the percentage of fares paid to them by Uber.

Furthermore, drivers are facing an increased competition from other ride-sharing services, further reducing their income. This, combined with the high costs of operating a vehicle and the unpredictable nature of the gig economy, has left many drivers struggling to make ends meet.

In the light of these challenges, many drivers are considering alternative employment options. However, this is easier said than done as the gig economy has created a culture of uncertainty and instability, making it challenging for drivers to transition to traditional employment.

The current conditions for Uber drivers are not sustainable, with many struggling to make a profit. The implementation of worker benefits has had a detrimental effect on their income, and the worker status had the opposite effect of what was intended. Uber needs to re-evaluate its business model and consider alternative approaches to ensure that drivers are adequately compensated for their work, or just switch to the older, working system.

33. The Future For Riders

The future of affordable private rides for Uber riders looks uncertain. The cost of private rides is likely to increase, and in the event that Uber crashes as a business, riders may be forced to book expensive taxis. Taxis are not subject to the same regulations as ride-sharing services, and therefore, riders may have to pay exorbitant prices to get around.

If Uber were to fail, riders may have to return to unreliable public transport, which may not be a suitable option for many. Public transport is often crowded and can be unpredictable, leading to a poor user experience. Furthermore, many riders have become accustomed to the convenience and affordability of ride-sharing services and may struggle to adapt to other modes of transport.

On the other hand, if riders choose to opt for taxis, they may have to contend with high prices and poor service quality. Taxis are not subject to the same competitive pressures as ride-sharing services and may charge higher fares as a result. Additionally, taxis may not have the same level of convenience or ease of use as ride-sharing services, making them an unattractive option for many riders.

Uber is now a vital business for taking people from A to B in any situation and place. When it will fail, the impact on private life and the economy will be terrific. This is valid for Uber's competition too.

Some of these companies are being used for large scale money laundering, and that may keep them afloat for a while, but that will only be temporary and it does not mean a good and reliable alternative.

34. Net Zero Vehicles

There are no fully green vehicles. People need to stop listening to politicians and interest groups promoting this agenda. Look at the science yourself! Electric vehicles are more dangerous to human health than classic "fossil" fuel ones, and that should not be overlooked. The best, closer to zero emissions vehicles or greenest at the moment are self-charging hybrids. Zero is not possible, it is just a marketing strategy the globalists are using to take control of your mind and life.

35. Overregulation Must Stop

Overregulation is often touted as a necessary tool to ensure safety and fairness in various industries, including the private hire industry. However, it can have a negative impact on both the industry and the drivers, particularly in the case of ride-sharing companies like Uber.

One of the primary issues with overregulation is that it can increase costs and decrease flexibility for drivers. For example, in some cities, regulations may require drivers to obtain special licenses or permits, which can be costly and time-consuming to obtain. In addition, regulations may require drivers to adhere to strict schedules or routes, which can limit their ability to take on more customers or adjust their schedules as needed.

Moreover, overregulation can also harm ride-sharing companies like Uber, by creating barriers to entry for new competitors. This can reduce competition, which can lead to higher prices and decreased innovation in the industry.

Furthermore, overregulation can also harm the customers by limiting their access to affordable transportation options. This is particularly true in areas where traditional taxi services are limited or prohibitively expensive, and ride-sharing companies such as Uber provide a more affordable and convenient alternative.

While some regulation is necessary to ensure safety and fairness in the private hire industry, overregulation can have negative consequences for both drivers and ride-sharing companies like Uber. It is important for regulators to draw a balance between protecting the interests of drivers and customers, while also allowing for innovation and competition in the industry. If regulation is hurting the industry, it is overregulation. Plus, it is anti-climate.

36. Blocking Anti-Individual, Anti-Freedom Laws For Good

In a democratic society, individual freedom is one of the most important values that needs to be protected. However, there are often legislative efforts made by governments that may be seen as anti-individual and anti-freedom. It is important for citizens to be aware of these efforts and to take steps to block them.

One of the primary reasons to block legislation that is anti-individual and anti-freedom, is that it can limit the ability of people to live their lives as they see fit. For example, if a government passes a law that restricts the freedom of speech, it can limit the ability of people to express their opinions and ideas freely. Similarly, if a government passes a law that restricts the freedom of assembly, it can limit the ability of people to gather and organise for social or political purposes.

Moreover, legislation that is anti-individual and anti-freedom can also harm economic growth and development. This is because it can limit the ability of individuals and businesses to innovate and create new products and services, which can ultimately harm the economy as a whole.

Therefore, it is important for people to be informed and to take steps to block legislation that is anti-individual and anti-freedom. This can involve speaking out against these laws, organising strategic legal actions, and contacting elected officials to express one's opposition. Ultimately, by working together to protect individual freedom, we can create a society that values personal autonomy and encourages innovation and growth.

37. Private Rides Are Greener Than Public Transport

In today's world, environmental sustainability is becoming increasingly important, and people are looking for ways to reduce their carbon footprint. While public transportation is often seen as a more environmentally friendly option, the reality is that private hire industry and private rides can actually be better for the environment in the majority of cases.

Firstly, public transport systems are not always as efficient as they seem. Buses and trains often run on a set schedule, regardless of how many people are using them at any given time. This means that they may end up running on empty or with only a few passengers on board, wasting valuable resources such as fuel and energy.

On the other hand, private hire industry and private rides allow for more efficient use of resources. With ride share services like Uber, people can share rides and reduce the number of cars on the road, thereby reducing carbon emissions. Additionally, many private hire companies are now using hybrid or electric vehicles, which are much more environmentally friendly than traditional diesel-powered buses.

Furthermore, private hire industry and private rides offer greater flexibility than public transport, which can often be unreliable or inconvenient. This means that people are more likely to use these services, which can reduce the demand for public transport and ultimately lead to fewer buses and trains running on the roads.

In conclusion, while public transport may seem like the more environmentally friendly option, it is not always the case. Private hire industry and private rides can actually be better for the environment in most cases, and it is important for people to consider all their options when it comes to reducing their carbon footprint. By promoting and using private hire industry and private rides, we can make a positive impact on the environment while still enjoying the convenience and flexibility that these services offer.

38. What You Can Do As Rider

If you are using the Uber platform, it means that you like it. This is why you need to take the same action as drivers do to protect this service, and not Uber, the company. If drivers are suffering, the service will collapse, so you must support the drivers' interests. You need to speak with your drivers and organise a legal siege aimed at TfL, Uber, the Supreme Court, your MP, your Council (drivers are local business), and even the Parliament, directly. Do not support union action or organise protests in London or any other cities. Protests are a good thing and a right we should never lose, but in this case it cannot change much, as the government will "spike" that protest with their undercover agents. Your action must be coordinated with others for quality action, but it needs to be individual. Centralised action with a leader will fail. Individual action coming from all directions will eventually put extreme pressure on regulators and our delegates (authorities), and will force positive change. The government likes centralised action because it finds it easy to corrupt and destabilise for the benefit of its anti-freedom, anti-individual "green" agenda.

39. What You Can Do As Driver

Firstly, stop joining groups that are promising to help you or your condition on the Uber platform. Educate yourself. It's free! Use this book to plan your action. Sharpen your language and mind, and adopt healthy progression. This means scaling up your legal action, starting with the smallest and closest player in the game. Approach Uber with requests, obtain answers (rejections), and then move to the next, up to The Supreme Court, and even the Parliament if your MP is being dismissive. Give them time to answer and make sure you get answers. If they are unapproachable, move on to the next level. Stay focused and even ask for professional legal advice. Just be careful with lawyers, as they might put you off by telling you nothing can be done and that being a worker is good for you. Only you know if it is good for you or not. No one else knows your condition. We are specialised in all types of law and we can tell you that there is a lot to do to reverse the worker status, and that it is quite simple with the right action and amount of individual legal actions. As of 2022, at least 5,000 drivers need to engage into legal actions, as per this book, in order to overcome the members of ADCU that have caused this destruction. ADCU is one action at any time, while 5,000 legal actions mean 4,999 more than ADCU. We do not think that when the lawsuit was filed against Uber by ADCU founders, they had the support of more than 100 drivers. The sooner you start your action, the better. Delay it and your suffering will be prolonged. This needs to change by 2025 or all may be lost, as probably Uber will fold its operations in the UK.

40. Why You Should Do Something

Whether you're a rider or driver, you must do something. Your wellbeing and freedom depend on it. If you look around you and try to see the big picture, you'll notice how the general conspiracy is taking us back to very dark ages, to deindustrialisation and technological regress. Uber was amazing progress that brought safety, affordability, and peace of mind to drivers and riders. Drivers could earn decent money, making them able to start other ventures (business diversity) and live healthy lives, while riders had decent rides at their disposal anywhere, and at affordable prices. The riders complaining about the surge prices at late hours in the night or when demand was really high, do not understand basic economics.

They should try driving at 2 a.m. with Uber sending them a £4.30 trip for 10 miles total drive. But they still booked with Uber, not taxis, which were probably double in price even with surge. People never complained publicly about taxi drivers charging extortionate amounts, but only about Uber. Transportation cannot be affordable at any time of day and in any conditions, as there are always people with more important things to attend to than others. This is how a free market works, but not anymore with Uber.

41. If You Don't Do Anything

In this scenario, all drivers partnered with Uber and others, will shift to normal jobs or doing something else. Losing private hire, this system that used to work, will have deeper implications than a simple mind can ever comprehend. As shown before, its importance was put to test during the 2020 pandemic. For that alone, it needs to thrive. There has never been a lower morale of Uber drivers since the start of the business in UK. If this continues many will get sick if they don't move on or take action.

Riders will lose affordable and instantly available private transportation. This will eventually have a negative impact on life. You will not be able to move as you please, you'll have to pay huge amounts to go places, spend more time on public transport, and lose considerable amounts of time to go from A to B. The implications are deep and may cause death to a lot of people. Uber happened and became successful, because there was a need for this type of service. Since you are now so used to it, just imagine losing it forever.

42. The Big Audit

What drivers must do, is ask Uber for a full audit. This is paramount to proving that the company has screwed drivers when it comes to pay (even before the worker status). We were willing to overlook these practices, as our income before 2020 was fair and substantial, but as we stand, and considering what Uber has done to us, they must comply to our requests for audit. We have proof that Uber was withholding money even from tips, as there were discrepancies in what we were shown in the app and what we got paid. The terms and conditions are not enough to safeguard the company against an audit. You might be told that they don't have to show you the full details of the trips, but they do. Drivers never see the receipts issued to riders after the trip. Since you have power over the services contract and because you're the one offering the service, they must comply. Plus, this is what being a partner driver means: full transparency. Uber is a Big Data company and they are recording very accurate details about everything you do on the app.

The audit must contain the following:
1. All receipts issued to riders (including secondary receipts for tips added after the trip has ended);

2. Time when going online/offline;
3. Time when trip was accepted;
4. Time taken to pickup;
5. Distance to pickup;
6. Price on the offer card;
7. Price at the end of trip;
8. Time spent on trip;
9. Time the trip ended;
10. All GPS data of the route taken.
11. Pickup/Dropoff locations;
12. Time of arrival for pickup;
13. Waiting time at pickup (many times not getting a cancellation fee despite waiting more than 5 minutes);

Uber will try to fight having audits for every driver with everything they've got, but if we manage to get it, we'll have them in the palm of our hand. There are 100% chances that they've paid us less that what they should've according to their own terms and conditions. This is a severe breach and liable to hefty compensation. Whatever they do, they cannot win against thousands of individual legal actions. With a full audit, it will be just a matter of time and formality until drivers get compensated and respected again for their loyalty toward the system Uber used to represent and entail.

You need to request an audit in writing from Uber's head office in London, not by contacting them through the app. Riders can help by liaising with their drivers and communicate the details of their receipts.

43. How Uber Planned To Cause Harm To Its Drivers With Clear Intent

In May-June 2021 Uber decided to organise a Q&A session aimed at promoting the unapproved mRNA injections to drivers across the UK. All this action was planned even with no casualties among Uber drivers after more than one year and a half into the pandemic, and after each driver carried hundreds of riders ill with the new virus and many others. Anybody can check these numbers. The percentage of Uber drivers that passed away due to this new virus is zero. And if zero out of the most exposed people out there died, then why did less exposed people die in such large numbers, only in hospitals? We know way more than we can expose here, for safety reasons, information we got from people working for the UK government, foreign governments, conspirators (secret and not so secret organisations and cults), and hospital staff. What we can say is that 100% of our riders that took these injections are feeling very ill. Imagine how desperate and scared these people must feel if they resorted to asking an Uber driver for medical advice. Uber's aim was to clearly promote the uptake of these injections, as no risks have been debated or mentioned throughout the session. This is why we believe that Uber's intention was to harm us. If you are to tell someone that you're planning to invest some money in Forex, straight away everyone will put you down by presenting you with the risks of such financial endeavours. When it comes to your health, apparently no one is willing to stop you from making a mistake. On the contrary, without any knowledge about the subject whatsoever, they will push you to take it.

Having extensive experience in immunotherapies, pharmacovigilance, and after studying the orange guide (the rule book in pharmaceuticals), we knew that this Q&A session will be interesting. One of the so-called medical professionals that were paid to attend is a famous TikToker. We could not wait to ask them questions that we knew they won't be able to answer. These questionable (one might say criminal) professionals are: Faith Uwadiae (immunologist), Karan Rangarajan (NHS Surgeon by day, full time TikToker), and Amir Khan (NHS GP). We are calling them questionable professionals because this is what they are. Their advice during the Q&A has damaged the health of Uber drivers that have taken the injections at their influence. All Uber drivers that have done it and spoke to us cannot recover completely from a simple seasonal cold. As a medical professional you cannot give advice on a good-for-all therapy without knowing the person's health condition (without an assessment) and everything there is to know about the therapy you're promoting. The biggest mistake you can make as a doctor is to listen to the government's advice, and tell your patients that they can take a certain type of medication that you know nothing about. These are no trustable doctors and we are praying for their patients. It is recommendable that these individuals are stripped of their honours and banned from the scientific world for life. As proof for their recklessness and incompetence, we'll present their answers (during the Q&A session) to questions asked by Uber, Uber drivers, and us. At first we did not believe that Uber will actually ask them our questions but they did, and we thank them for that. Little did these rogue impostors know that on the other side there were true professionals too, trying to squeeze real answers from them and uncover their preposterous intent. We have succeeded as you will soon find out. Let this be proof that you should never trust anyone but yourself when taking any pharmaceutical product. Do your own research, learn medicines and protect yourself and your family from criminal minds and intentions. By the way, these injections are not free. We are all paying for them and our children will keep on paying for them in the following decades. Do you still care about the others? Isn't that ignorance and selfishness?

    Based on the racial profile of the guests you can clearly tell that they were targeting Uber drivers in a subliminal way, to influence as many as possible to take up the injections. The majority of drivers are black, South Asian, or of Arab descent. They didn't bother with targeting white drivers, as we are a very small minority on the Uber platform. How they approached the session, by targeting racial profiles of drivers, is criminal in itself.

The following is the exact transcript of the session, minus the irrelevant chatter.

Question no 1 (asked by Uber or another driver):
What is the vaccine and how does it protect me from COVID-19?
Faith Uwadiae (immunologist) responds:
"Yeah. So basically a vaccine it is a type of medicine and the key thing about it is it's about producing an immune response that will protect you from getting a disease. So it can target a number of different infectious agents so it can be bacteria, can be viruses, fungus or parasite. But the key thing about how that actually works is that it educates your body on how to actually react without you having to have a disease or risk, any of the complications that come with it. So, they generally work by introducing some sort of part of the virus or instructions for your body to make a virus, for example, and the key bit is that you could produce an immune response which can be antibodies that target it or it can activate cells in your immune system to target it and basically educate your body on how to react to it, again, without you having to actually go through any of the risk that you have in getting a disease. So it's a safe and preventative measure of educating your immune system."

Conclusion:

As an immunologist, she's giving a generic answer about how vaccines work, more or less, and does not touch the topic of mRNA technology present in these injections, and what they do to your body. Her answer is just promoting the mRNA jabs by avoiding the answer. The mRNA jabs are nothing from what she said above. None of her words touched the implied topic of the question. The author of the question clearly referred to vaccines anti COVID-19, and since all injections that were produced contained only the mRNA technology, it is implied that the question was referring to this. The immunologist replied with the definition of vaccines, with grave scientific errors, as no vaccine, whether it works or not in producing immunity in the body, it can never protect you from catching and getting sick with the real virus. If you do not get infected with the real virus, you cannot develop the right level of immunity. It can be partial, but never at full potential with a vaccine. It would have to contain the real unattenuated version of the virus, meaning getting infected intentionally. Ms Uwadiae is either not an immunologist or very uneducated on the topic of her career. For what she said during the session and what we know in 2023 about the harm caused by this mRNA technology, she should be banned for life from practicing in the scientific world, because she is a menace to human life. We have never seen such irresponsible behaviour from medical professionals before 2020. We had the trump card over the other drivers for having worked in immunotherapies (extracting antibodies from human blood plasma), but even so, knowing more about immunity than a trained immunologist is very dangerous for the general population. This is why we wanted to ask them questions that might stimulate the drivers participating in the session to think it through the right, scientific way.

Question no 2 (asked by Uber or another driver):
What is in the vaccine?
Faith Uwadiae (immunologist) responds:
"Okay. Yeah. So vaccines have a number of ingredients, and it varies depending on the type of vaccine with the method that's used to produce that immune response. But if we're talking about vaccines in general, the main ingredient's actually water, which I think that people don't really think about so much, but it also has some active ingredients. So, for example, the main active ingredient is what you're going to use to generate your immune response. So, we take the COVID-19 vaccines, for example, you have the Pfizer-Biontech vaccine, which the mRNA vaccines, that's to provide RNA to your body, which kind of gets it to, um, give instructions to your bodies to produce the spike protein of the virus, which is on the surface, and your body reacts to that. And you also need some sort of method of delivering the vaccine into your bodies. So that can be in the case of

the Pfizer-Biontech vaccine, some sort of fat that surrounds the messenger RNA. And then you have a number of stabilisers, such as sort of water or things that preserve it, so that it's easy to transport. But again, I said the main ingredient is essentially water, but you want to have something to provide that active ingredient, some form of delivering it and make stuff to make it easy to store and easy to transport as well."

Conclusion:

Again, as an immunologist, she gave a fifth grade answer, creating an even bigger dilemma for the attendees. Vaccines do contain water, which is called "Water for Injection", the purest and safest water on the planet, with zero conductivity (it does not conduct electrical current) and a purity of less than 0.4 ppm (parts per million). They also contain immunity stimulators, that basically harass and force your immune system into producing immunity to something you shouldn't produce (waste of resources), patented technology (you will never know what it is), the lipids containing the mRNA instruction (gets absorbed by your cells that will start producing proteins you should not produce – the immune system will take them out), and other products that are not supposed to enter your body. We are referring here strictly to mRNA injections, which should never be called vaccines, as they don't fit the definition. They are gene therapy. Everything that is changing the natural behaviour of your cells is gene therapy, and a trained immunologist should know this. The mRNA jabs are changing the chemistry of your body, your natural expression. Since your personality is being dictated by your unique chemistry, just imagine what this means for and it does to you. Ms Uwadiae must to be stopped from practicing medicine for her hazardous lack of judgement and skill.

Question no 3 (asked by Uber or another driver):
Is the vaccine effective? What if I have immunity?
Karan Rangarajan (NHS surgeon) responds:
"Yeah, so you know the vaccine is very effective. There've been lots of trials done by all the various manufacturers and, you know, the trials have included, tens of thousands of people. And we've also got real world data from real people who've had the vaccine and you know, we've had really good results and it's reduced deaths. It's reduced the number of people going to hospital or suffering complications from it. So it's very effective based on the trial data we have from all the studies and from The Real World data as well. So we know it's very effective and will

stop you getting COVID, and you know, definitely stop you getting very severely unwell. And you know, even if you've had COVID before, I've had COVID-19 last year. It's still, you know, important to get the vaccine because especially with the new variants you can still get COVID, and just having COVID and recovering from it, doesn't guarantee lifelong or a very long immunity. You might just have a few weeks of protection. Getting the vaccine gives you longer lasting and stronger protection. Um, you know, if your body learns to build antibodies from a vaccine that, you know, gives you long lost immunity. So, that's what I would say."

Faith Uwadiae (immunologist) intervenes after being asked by the host if she's got something to add:
"No, I actually thought it was covered very, very well."

Conclusion:
Karan is very famous on TikTok. An NHS surgeon and social media buff, with patients waiting on very long lists to get adequate treatment. Today, for blood clots too. We have studied these trials from the beginning and our experience in immunotherapies told us to search for something very important, for antibodies. Because when all these trials started, most people already had the new virus at least once, hence they all should've had ongoing strong immunity. The

immunity for a coronavirus variant lasts for years, not weeks as Mr Karan is saying. He is so wrong that we feel very sorry for his patients. The immunity for all coronaviruses lasts roughly the same. You never lose the immunity for one coronavirus faster than for another one. If the immunity for a coronavirus would deplete in weeks, then we would always test like immunodeficient people, with very low levels of antibodies, or at least the normal reference levels in testing laboratories around the world would be lower than they are. In addition to this, due to amazing contributors in immunity, as the dendritic cells (dormant memory immune cells – they hold a record, a ledger of past infections) and macrophages, your immunity for a coronavirus and its multitude of recent mutations is for life. You only get sick with the same coronavirus once your antibodies for that virus have depleted (it'll turn into a normal cold), or when it modifies so much that it becomes a totally different coronavirus altogether (usually in people with autoimmune diseases). This can take years. As a healthy person you never get sick twice in a year with the same coronavirus. All coronaviruses that can infect humans are over 85% the same, and that is something your body can recognise. Mr Karan is totally oblivious to this. When we looked at the trials data, we could notice with ease that nobody tested the participants for antibodies against COVID-19. They were only tested with the PCR test, which is not a diagnosis method, nor for testing immunity. You'd need a specific IgM/IgG test for that. They were never performed. All these participants in the trials came from the real world and they were sent back into the real world after the mRNA shots. Considering that many of them received the placebo (saline solution – water and salt), it was easy for the pharmaceutical companies to claim that the naturally developed immunity was a cause of the jabs, in a high percentage. If you have had the virus once, you do not need any medicine, as that can severely destabilise your immune system. Healing is holistic in most part, and interfering with it recklessly can completely alter your healing capacity. What we don't need is more people with immunodeficiencies. Playing with logic a little bit as well, if you cannot build strong immunity by having the virus and recovering from it naturally, how can your body produce strong immunity from a vaccine? That has never happened in our entire medical history. Immunity can only be supported, not improved. The moment you try to improve it, you will destroy it. Just remember that wild animals have never been vaccinated, and they are thriving. If human genes cannot transmit good immune abilities, then human kind will disappear sooner than later, and caused by us, by people like Mr Karan.

Question no 4 (asked by Uber or another driver):
Are the vaccines protective against all variants?
Karan Rangarajan (NHS surgeon) responds:
"Yes. So you know, as you may all know, looking at the news, there are so many different variants cropping up all over the world. You know, South

African, the Kent strain, the Indian variant now, which is of concern. But you know, the data we have at the moment and, you know, you have to understand that, you know, coronavirus is a new thing, so we constantly pick up new data, but from the data we have right now, the vaccines protect against all the variants and we have actually a recent study done in the UK which shows actually that two doses of the vaccine will give you very, very good protection against the Indian variant, 88% effectiveness with two doses of the vaccine against the Indian variants. The vaccines do help. Even if there are new strains, the best way to protect yourself apart from wearing masks, social distancing, and washing your hands, and ventilation, is getting the vaccination. So, you know, get the vaccinations and protect yourself is what I would say."

Conclusion:

So, a surgeon is telling people to look at the news to see how new variants of the coronavirus are emerging (i.e. Matt Hancock WhatsApp messages). As we are writing this, in 2023, after all the data released forcibly by the pharmaceutical companies producing the mRNA jabs, considering what we know, Mr Karan is a clear danger to human life, a menace to the real medical society, as well as society as a whole by any definition, and should immediately be put under investigation for manipulating and influencing people with his erroneous advice. As a doctor you have a tremendous responsibility to protect people from malicious medicine and misinformation. How he replied to this question is horrifying and detrimental to his career. We are not saying that he never did good, but accepting to be paid by Uber to push the mRNA injections uptake among Uber drivers is a serious crime that should result with him being stripped of his honours, and at least banned from the medical profession. He's not a dumb guy, as you can see from his TikTok posts, but what he has done during this pandemic has completely erased his good history. We are not saying that he does not deserve mercy, everybody does, but this type of professionals have to be reassessed, to stop them from causing more harm in the future. With his advice during the Uber session, Mr Karan literally killed people. He offered no professional explanation as to why drivers should take the jabs, despite knowing very well how dangerous the mRNA technology is. You never promote a type of medication without discussing the side effects first. And with the mRNA technology there are terrible repercussions for the human body.

Question no 5 (asked by Uber or another driver):
What about the Indian variant - how dangerous is this variant? How can we protect ourselves from

it?

Amir Khan (NHS GP) responds:

"Yeah, so this has been in the news quite a lot over the last few weeks. What we know about the Indian variant, also known as B1617 variant, has different kind of mutations within that variant as well. Um, but what we know about it is that it is likely to be more transmissible because of one of the types of mutation in it on the spike protein, of the, of the virus. That's the part of the virus that attaches to human cells and mixes, fuses with them almost, and allows the virus to, to get into human cells. So it can be more transmissible. But there's no evidence to suggest that it actually produces more serious disease, then, then let's say, the dominant strain that we have here, the UK variant. Um, how can we protect ourselves? We've just heard that vaccines are, are really effective against the Indian variant and that that has been shown in these UK studies. But there's also some US studies to support that. Ah, so the best way to protect ourselves, as, as we've just heard is wearing a mask, maintaining space, washing your hands and improving ventilation in indoor spaces, including cars, ah… And, and getting both doses of the vaccine, and that is key, it has to be both doses of the vaccine."

Conclusion:

What we know from talking to many GPs across London, as riders, we found out that they know nothing about viruses and vaccines. What they do know is from the pharmaceutical companies and the general approved narrative of the NHS. We were shocked that they do not know how the factor-D vaccine, that we were producing in Bio Products Laboratory, actually works. GPs are family planners and they don't know how important it is for a new mother to know her blood's Rh. We were explaining this science to GPs! And many of them with decades of experience. So, in this case, asking a GP about taking an injection or not is a big mistake. They are not trained in the science of vaccines, nor medicines in general. As proof, he is saying that a mutation of the virus means that the spike protein is mutating, not the RNA sequence of the virus. A new variant is a mutation of the RNA sequence of the virus. Looking at the spike protein alone you will have to take an injection on a daily basis, as there are millions of variants of the new virus around the word. That doesn't mean that you'll get sick with all of them. That will never happen, as having been sick with the original virus gives you lasting protection against all variants. We are clear proof of that, having had the new virus once, in February 2020, and being completely healthy for two years after interacting with ill people with sickness caused probably by all variants around the world. A GP will never, and by agreement, is not allowed to speak against any vaccine, even if we know that the swine flu vaccine killed millions around the world. If they do that, they will lose their licence to practice. Trusting a GP when it comes to medication is one of the biggest mistakes you can make as patient. You need to do your own research about everything that enters your body. Yes, you need to read medical studies and the medicine's official documentations, including the trials data. In this case, how many of you know that many from the mRNA trials have died, some could not be contacted anymore, and that the real data about how the efficacy had been calculated was never made public. As said in the beginning, they never tested people in the trials for immunity, which rendered the trials completely flawed. These mRNA jabs do not give you immunity to any virus, confirmed by a mountain of data as of 2023. Mr Khan looks and sounds like an impostor, and has to be stopped from giving any medical advice. People like him are killing patients.

Question no 6 (asked by Uber or another driver):
Why does this virus keep mutating?
Amir Khan (NHS GP) responds:

"Yes. So viruses mutate. It's part of their natural life cycle. So when they infect someone, the way they work is, they will bind to human cells, get inside and program that cell to produce more of the virus, which then go on to infect other cells within that person. And then that person then spreads the, the, the virus by coughing, breathing, that kind of thing. Now, the more the virus is allowed to spread the more cells, it will enter and the more replication it will do. So it'll keep producing more and more copies of itself. As with anything that is a natural process, errors occur. So the virus replicates, replicates, within these numerous replications errors will occur and these errors are known as mutations. The vast majority of mutations are harmful or ineffective to the virus and are not carried on. But every so often due to chance a useful mutation to the virus will occur. That may make it more transmissible more…might make the disease more severe. And if the virus is more transmissible, it will quickly overtake the dominant variant in that area because it can spread quicker. And so that one becomes more dominant. The best way to stop viruses mutating is to stop the spread of them. The more people they have to spread in the more likely they are to mutate. So, we can cut that kind of chain of events through social distancing and through vaccinating people. We can reduce the number of mutations that

occur, but like I say it is natural for viruses to mutate, but we can, we do have it within our power to break that chain of events and reduce the number of mutations occurring."

Conclusion:

Again, Mr Khan gave a kindergarten answer to a very simple but important question. We don't think most people didn't know how viruses spread, but his lack of scientific knowledge is outstanding. Viruses don't run after humans at various speeds, in order to be more transmissible. If it were so, we would have to "eat" vaccines for breakfast, lunch, and dinner. A completely new variant will never form so often, to have a new one every week or so. And it does not depend of the virus but on the infected person's ability to develop a robust immune response to fight the infection. That is why immunocompromised people have to protect themselves, isolate until the rest have developed herd immunity. No one knows who's immunocompromised and why, so it's their responsibility to stay away from others around them. Since this virus can be passed away by animals too, like bats, which are everywhere, self-distancing if you are healthy is futile, and it can only lead to the decay of human interaction around the world. If a coronavirus is out of a lab, you will eventually catch it. Your wellbeing depends on how you manage your illness, the support therapies you are receiving. Mr Khan's crime becomes even more grave as he is saying that all people should get these mRNA jabs, without any assessment of the person. For example, if you put these jabs into immunocompromised people, you will kill them sooner than later, as with RNA vaccines (flu jab, etc.). All injections contain substances that people are or might be allergic to, and once in the system they cannot be taken out. Promoting such practices without evaluating the injections or the personal circumstances of patients is a serious crime. Malpractice is a leading cause of death around the world. If Mr Khan would've promoted this harmful gene therapy and hurt patients during communism in Eastern Europe, he would've been sent to prison for many years, if not sentenced to death by existing laws. It is absolutely shocking that in a presumed democracy nothing is happening to these professionals, no matter how they behave and how many people they are hurting. But we truly believe that a time will come when they will have to face their actions in a court of law. Too many people are dying because of people like Mr Khan. If you paid attention to what he said, that *"the vast majority of mutations are harmful or ineffective to the virus and are not carried on"*, you'd understand that he makes no sense whatsoever. He probably meant harmless? This coronavirus keeps mutating because scientists are looking at it. It can even mutate in a sample of saliva outside of the body, right before it gets analysed. If you test 10 different people, ill with the same coronavirus, the chances of finding at least one new variant are really high. All coronaviruses mutate, but it'll take years before you get sick with the same coronavirus again, because all variants are over 99% the same virus for which you have a memory of antibodies (inside dendritic cells). And viruses mutate rapidly in immunocompromised people. If you stay away from them you will not get sick again very soon. Immunocompromised people are the most dangerous spreaders around you, because their bodies are struggling to build immunity, therefore giving plenty of time to the virus to mutate in a considerable way. The mutation sits in the RNA sequence of the virus, not in the spike protein. The RNA dictates the proteins on the outer shell of the virus cell. Even if the spike protein used to penetrate human cells does mutate, it makes absolutely no difference in terms of infectiveness. Having developed immunity against a virus, even if a new variant penetrates your cells, you will not get sick for years, as the remaining antibodies will take out the infected cells and the virus cells no matter the variant.

Looking only at antibodies is a flawed way of looking at immunity, as there are many other barriers before antibodies get produced. At most, you'll experience cold-like symptoms for a brief period of time, maybe a couple of days. When your antibodies get depleted, your dendritic cells will keep the memory of the right chemistry of the virus, and the new infection will be handled in a very swift manner, as they will communicate with the immune B cells via antibody-antigen molecules in record time to produce the right type of antibodies. This is why you don't need any vaccine if you are healthy. As proof, all MPs in the UK, or at least most of them, have never had the mRNA injections, and they are still around and healthy. If they show up in the NHS system that they did, they paid for that registration. We know that as a fact. Based on our experience with riders that took these jabs, it is impossible to have no serious side effects from the mRNA technology, if you weren't lucky enough to get the saline solution. But we doubt that anyone can be so lucky to get only saline solution with 4-5 injections. If someone is telling you that they had all jabs and felt nothing, they are lying to you.

Question no 7 (asked by Uber or another driver):
Is the vaccine safe? What are the short term & long term side effects?

Karan Rangarajan (NHS surgeon) responds:

"Yeah, so again, going back to, you know, two parts of this question, is the vaccine safe? We know it's safe because there have been lots of trials done on the vaccine, and the vaccines wouldn't be approved for use unless they were proven to be safe with no significant side effects. And so, you know, in the tens of thousands of people who have undergone on these various trials, there have been no red flags at all. So you know, the various quality assurance board and the governments of different countries have said they're safe to use, and in the UK we have quality assurance process, the MHRA and they're really, really strict on what medicines and drugs are allowed to be used. They're one of the toughest boards in the world, it's one of the, you know, world leading, you know, quality assurance processes. And they've said the vaccines are very safe to use. And again now we have real-world data, you know, millions of people across the world have received vaccines, with no significant adverse events, that's you know, part 1 of the question and in terms of short-term and long-term effects, what you need to know about vaccines is that generally the vaccine doesn't stay in your body longer than a few days at most. Um, you know, it just it goes away, it doesn't cause typically long-term effects, usually you get some very mild short term effects that

don't last longer than, you know, maybe a few hours or a few days at most and the type of stuff that you typically may experience, and not everyone, is things like pain where you get the injection in your arm, it might be sore for a few hours you might get a mild headache, some general muscle aches or mild temperature, you know, just general aches and pains in your body. But again, you know, I've had the vaccine, I've had both my doses, and after both doses I experienced a little bit of a headache and muscle ache for a couple of days and then I was fine. And that's what, you know, a lot of people do experience if they get the side effects, but you may not get anything at all and you may be absolutely fine. And the chances of something slightly more serious than that is one in several million, you know, you've got more chances of being struck by lightning, you know, than having a serious effect from a vaccine and you protect yourself against COVID and we know COVID can have serious short-term and long-term effects, so the vaccine is much safer to do that."

Conclusion:

Everything Mr Karan has said is wrong. Again, he did not touch the subject of what the mRNA technology is doing to your body, that it reprograms your cells to produce a protein that they should not produce, and that if that happens with pancreatic, liver, heart, neurone, or kidney cells you will die sooner than later, even from one dose. Which actually is happening all around us, as we are writing this book. We have had riders that experienced the collapse of both their kidneys overnight due to these injections, despite being completely healthy before that. We even had medical professionals that got fooled into taking the jabs, and who are now living with painful arthritis in their joints among others. The trials for the mRNA injections have had massive red flags from the start, with people dying shortly after receiving them, but they were not being reported publicly. The MHRA knew about everything, but because they are being funded by the Big Pharma, they did not try to stop anything. We have seen how the MHRA are regulating and it is not meant to protect human health. Plus, people at the top of the regulatory body are being named in those positions by the Big Pharma. There were hundreds of thousands of people participating in these mRNA trials, despite being known that it is not a therapy for humans, but no regulator protected them from harm. This is the answer for you about how regulators operate. Mr Karan is referring to RNA vaccines, but even there he is very wrong. And it is not about the fact that the jab stays inside of you for short or long time, it is about what happens with your body once you start producing massive quantities of a protein that is poison for your body, process which can last for years, based on real scientific data. That protein will destroy billions of your cells in that timespan. Some cells regenerate, but other don't. When you lose them, you lose them for good, and there is no replacement. What happens then, Mr Karan? How do you replace neurones, pancreatic cells, heart cells, kidney cells, etc.? These vaccines were never approved. EUA (Emergency Use Authorisation) does not equal, nor mean full approval. The EUA was brought in to give people something for a lot of money. The MHRA and the FDA went against all pharmacovigilance protocols when pushing for these injections. And so, they are committing genocide around the world. When they released this gene therapy they knew exactly what the short and long term side effects will be.

Question no 8 (asked by the authors of this book):
What changed in the mRNA technology since it was categorised as highly dangerous by the same regulators?
Karan Rangarajan (NHS surgeon) responds:
"Yeah, so as far as I'm aware there's no categorisation by the regulators that it was dangerous, and mRNA, it's, you know, it sounds like a new thing that we've just, you know, scientists have developed over the last few months, but actually if you look into the history of mRNA technology, it's been developed and researched since the 1960s, so actually there's been about 60 years worth of research into mRNA technology, so it's not new, it's you know, tried and tested in other things but it's a very, you know, well established thing, and it's now been

transitioned and used into helping us to combat the virus, it's a new technology that can be rolled out faster and it can be changed to combat new strains quickly as well. That's the advantage of mRNA technology. It's not dangerous. The reason people think it's dangerous is because they confuse the mRNA with DNA and think that it has some sort of interaction. The DNA is inside the nucleus of the cell, which is completely different to the mRNA, ah, so actually it's not dangerous and it's very well researched technology which we're now using, you know, for a different use, for vaccines."

Conclusion:
Mr Karan is again completely oblivious toward the scientific research into mRNA technology. When saying that mRNA technology has been studied since late 60s, he means actually RNA. They started studying and producing new sequences of RNA around 1960s, when they also found out that alien, non-human RNA can fuse with human DNA, as over 50% of DNA is identical to RNA. Scientists might have played with mRNA lipids before but the technology was fully understood and modernised around 1986 by Dr Robert Malone, who is also experiencing severe side effects of these new mRNA injections. So, a pioneer of this technology is against it being used in humans. How can people as these panelists be more experienced in the mRNA topic than Dr Robert Malone, THE expert in this domain? And yes, Moderna, one of the companies producing

mRNA vaccines has deemed this technology very toxic for humans, when they were experimenting with it, around 2010. They even scraped the research into it for a short while. Nothing changed in this technology along the years. What they did starting 2019, is to push for developing capabilities to produce jabs for the whole planet. The mRNA technology stayed the same. And in under no circumstance is it cheaper to produce compared to normal RNA vaccines. It is a complete lie, as it is way harder to isolate each new RNA sequence to put in the mRNA lipids, plus synthesising vector rhinoviruses to carry the RNA. Whereas with the flu jab, they're just using 3-4 main strains circulating at any given time, they purify them (making them inactive), and they inject them in the people aided by other patented technologies. During this pandemic, they were not allowed to use the real virus because it is not a natural occurrence, it is patented by the lab in Wuhan or someone sponsoring their research. They did not allow the use of the real virus because there would've been huge lawsuits between the jab manufacturers and the Wuhan lab for patent infringement. That would've uncovered the whole conspiracy and illegal research. Even so, the truth is finally coming out. We're just wondering what the Uber panelists are thinking now, after destroying the lives of thousands of people with their uninformed and malicious advice.

Question no 9 (asked by Uber or another driver):
How long does the vaccine last? Will COVID ever

be eradicated or will we have to get a vaccine every year?

Faith Uwadiae (immunologist) responds:

"Oh, yeah. So, if you mean by how long does the vaccine last, it means, like, the protection from the very illness. So, if you're given the vaccine, we know that antibodies at least seem to last for at least six months, and that should provide some sort of protection. In terms of how long after that lasts, that's ongoing, and the reset will eventually tell us that, but it should provide a good, long lasting protection up to at least six months, we know. In terms of whether or not COVID will be eradicated? I think that's a really interesting question because, I think, when we think of the concept of herd immunity, so if we can vaccinate enough people in a country, for example, so that you can try and protect people who wouldn't be able to take a vaccine or people who would be too young to take a vaccine, um, in terms of whether or not we can actually reach herd immunity is quite hard to assess. But what we do know is there's a possibility that essentially we could have to live with it for a long time. So, it might be that it ends up being like the flu vaccine, where we'll have to constantly top up the vaccinations so that we can tackle current strains that are there, such as variants, and we'll just top them up and then just protect the most vulnerable in our society, for example. But we might be able to eradicate COVID maybe from some parts of the world, but whether or not we can eradicate from the whole world, I think only time will be able to tell us that, overall."

Conclusion:

Ms Uwadiae does not know that you cannot eradicate a virus, less so with a vaccine, in this case with a gene therapy. Even the plague is still with us, killing roughly 100,000 people around the globe each year, usually in very poor countries where rats roam freely in the community, in places such as India. The unscientific topping up your immunity theory coming from an immunologist is utterly shocking. She basically does not know how immunity works, and all she's doing is promoting the mRNA technology. She should be ashamed of herself, as she is talking as a vaccinologist working for a pharmaceutical company. There is no science proving that these vaccines are giving you immunity to any coronavirus. That is not even possible, because you need to interact with the entirety of the virus, not just a single spike protein, in order to develop a real immune response. Even the flu jab does not give you immunity to the real virus, as you're getting an attenuated version of it. The aim there is to not feel really sick when getting the real flu, as you will get infected with it, usually every 10 years (give or take) in our experience. Every time you get the flu jab, you're wasting important resources, because if you've ever had the flu, you have developed immunity for life, no matter the strain you get infected with later. All that matters is to live a healthy life and what you do when you get sick, so you recover really quick. If you smoke and eat from McDonald's every day, then you should not be considered when declaring a pandemic, because if you don't care about your health, what do you care if you're vulnerable when contracting a respiratory virus? Only how healthy people react to a new coronavirus should be considered as the standard in declaring a pandemic. If no healthy person dies from the virus, or a very low, insignificant percentage are dying, then there is no pandemic. The rest of the population should only receive immunoglobulin shots with naturally developed antibodies from people that had it. That should've been the only solution, with no need to pump billions into new logistics and facilities, as they are already there for the vulnerable. Bio Products Laboratory can produce immunoglobulins for all vulnerable people in the world in less than six months. Now, the vaccinated cannot save the immunocompromised anymore, because the immunity developed from vaccines, less so from gene therapies, cannot be shared via blood plasma donations. The flu jab immunity is short lived, contrary to the naturally developed one, by getting infected with the real virus. Ms Uwadiae does not know anything about immunity. She was just paid by Uber to promote the mRNA jabs to the drivers on the platform. This is a very serious offence from both Uber and the panelists.

Question no 10 (asked by Uber or another driver):
Can you still catch or transmit COVID after having the vaccine?
Mr Karan Rangarajan (NHS surgeon) responds:
"Yeah, so, you know, you can still catch COVID after

having the vaccine. It's, um, you know, not every medication just like the simple, you know, pain relief or anything you buy over the counter, it's not 100% effective. Nothing is 100% guaranteed. But the vaccines are very, very, very highly effective. And but unfortunately, there will be a very, very, very small percentage of people for whom the vaccine won't work, which is why we always tell people to continue wearing masks and washing their hands and social distancing even after getting the vaccine. So you can still catch it. It's unlikely and it's quite rare, but it does happen in a small percentage of people. But even, and this is the really key thing here. Even if you do catch COVID after getting the vaccine, the whole point and the main purpose of the vaccine is not actually to stop you getting COVID, it's to stop you getting a really bad COVID that's going to put you in the hospital. That's the actual number one aim of the vaccine, is to make COVID almost harmless, just like the flu. Number two is stopping you getting it completely. So actually, that's one thing a lot of people don't realise. So you can if you get it. You just get a runny nose or a sore throat or a bit of a headache, and you actually are saved from getting a bad one that lands you in hospital or worse. There is, you know, limited data to suggest that getting the vaccine reduces the spread as well. It might limit how much of the COVID virus particles you spread around, so it's actually, you know, good protecting yourself and other people as well."

Conclusion:
Even with this question, Mr Karan is failing to explain how the mRNA can provide you with a potent immune response against COVID, or if it's ok for your body to produce poison for long periods of time. His lack of judgement is terrifying. At the time of the presentation there was enough concerning data about these jabs for him to pull back his advice. Yet, for big money, he decided to contribute even more with killing people.

Question no 11 (asked by Uber or another driver):
Am I safe after the first dose?
Ms Faith Uwadiae (immunologist) responds:
"Yeah, so if you have one dose of the vaccine it does provide some level of protection you start getting antibodies and cells of the immune system that will help you fight off infection, but really, you're not truly safe, so you have two doses of vaccine because the second dose is really there to boost that initial immune response that you have to a level that can provide a good level of protection, so really two doses, but there is some protection after one dose."

Conclusion:
What she is saying above is to confirm that the mRNA technology does not work, and more than that, that the flu jabs never worked either. If this dangerous technology is so amazing in giving you immunity against COVID, and no other technology is better, then normal RNA vaccines have never worked. When

you contract the real virus you don't have to contract it again in three months time in order to become "more" immune to it. You have it once, and then you have it again in a few years time, next year, it all depends on how you live and how healthy you are. Ms Uwadiae's aberrations are gross and very dangerous. She fails again in explaining the mRNA technology. It is truly outrageous that we know more about this than her, just by studying official documents and the science behind it. If the biggest schools in this world that are training immunologists do not know this science, then what are they training? Killers?

Question no 12 (asked by Uber or another driver):
Which vaccine should I take and can I choose? What are the differences? Which is the safest?
Mr Amir Khan (NHS GP) responds:
"Yeah, I'm gonna break this one down, so which vaccine should I take and can I choose? The guidelines in the UK for anyone under the age of 40, um, I'll give him the mRNA vaccine, so that is the Pfizer or the Moderna. Anyone over the age of 40 is offered whatever is available in your vaccine centre at that point, so that could be Oxford-AstraZeneca vaccine, and that is because we know that the benefits of having any of the three vaccines that are approved here in the UK, the Pfizer, Moderna, Oxford-AstraZeneca vaccine way outweighed the risks, um… that come with the vaccines when you're 40 and over, so you can have all three, any of the three, but it's

important that you get the same vaccine for both doses. If you're under 40, even though the risk is incredibly small, incredibly small, and there hasn't been a definite association between the Oxford-AstraZeneca vaccine and the risk of these very, very, very rare blood clots. We advise people under 40 go for the Pfizer or the Moderna vaccine. Um, now, can you choose? No, not really. If you're over 40, really the best vaccine is the one that is offered to you, um, and if you're under 40 you will only be offered the Pfizer or the Moderna one, depending on which one they have in stock, so please, please, please take whichever one you're being offered. There is no point delaying it for any reasons, because the more you delay it, the longer the risks are to you. So what are the differences? I think I've explained a little about them there, um, but they are, in how they work, so the Pfizer and the Moderna ones are these mRNA vaccines that Ms Faith mentioned, and the Oxford-AstraZeneca is a weakened virus, genetically modified to contain parts of the coronavirus that won't make you sick, but will give you a good immune reaction, should you come across the real coronavirus and you don't get sick for it. Which is the safest? They're all safe, and the reason why they're age stratified like that, depending onto risk benefit, we offer the safest one for you, but in general terms they are all safe, and you will be offered the ones that are safest for you."

Conclusion:

As you've probably noticed, Mr Khan is talking about levels of safety. When someone is talking like that, it means that nothing is safe. In 2023 they're giving any combinations of jabs to any age, and because of that you can never know which is more dangerous than the other, as it is impossible to find the riskiest when people have taken a combination. Safe is not healthy, and GPs do not know anything about vaccines and immunology. They are only pushing forward the procedures and policies, as well as the science dictated by the NHS, and since Big Pharma controls the narrative of centralised medical services worldwide, it is their view that's being pushed forward. In private clinics you might find good and compassionate doctors, but even there it is very rare. All these jabs contain particles that enter your cells and wipe out millions, if not billions of them in time. What Mr Khan said above makes absolutely no scientific sense, and in 2023, when many young adults are dying "suddenly", his words are a crime against humanity.

Question no 13 (asked by Uber or another driver):
Can I take one type of vaccine for my first dose (eg. Pfizer) and a second type for my second dose (eg. AstraZeneca)?
Mr Amir Khan (NHS GP) responds:
"So there is research going into this at the moment to see whether mixing doses has any effect or even a beneficial effect to people's immune response but we don't have the data available yet, so as things stand, if

you get the Pfizer vaccine for your first dose, you must get that for your second dose. And that applies to all the vaccines, um, so at the moment whatever you get for the first dose you will get for the second dose, and that's the only one you'll be offered, so, yeah, stick to what you're offered."

Conclusion:
As with previous answers, Mr Khan knows about vaccines and immunology what is being served from the central medical system, the NHS. As we are writing this, after 12 February 2023, it is no longer possible to get the initial COVID jabs in vaccination centres, but only boosters. As things heat up, with people dying at a rapid rate due to this gene therapy, the NHS is slowly and silently closing all vaccination centres across the UK.

Question no 14 (asked by Uber or another driver):
How long did the vaccine take to develop and what was the process? Are we still in the trial stage? Why was it developed so much quicker than others?
Mr Karan Rangarajan (NHS surgeon) responds:
"Yeah, so you know how long does the vaccine take to develop? Obviously we know that coronavirus struck us towards the back-end of 2019, December-January, that kind of time, and the trials, um, the first trials Pfizer trials, which started in July 2020 and then, um, you know, the first person in the world to get the Pfizer jab was a 90 year old woman, um, in the UK,

so there's a 6 month gap between the stages and the vaccine actually coming into the arm of a person, so about 6 months, um, so that was the trial stages and it was approved and, you know, to people in the UK and the rest of the world as well, so we're not in the trial stages, the trials have been done extensively tested on people, no significant side effects, no safety concerns, so we're not guinea pigs in some big trial, that is the most important thing you need to know. There will be trials which follow us up, so people who have the vaccine might be followed up just to see, you know, in terms of how long immunity lasts and things like that. But we're not in an experiment at the moment. That's the key thing you need to know. And another common question like we've got here, why was it done quicker than all the other vaccines in history? One key thing you need to know, the actual development stage and the testing stage of the vaccine took the same amount of time as vaccines of previous years as well. The only thing that was quicker was all the funding and all the actual, you know, the research approval, you know, the trials need to go through ethical approval and all this kind of stuff, so all the administrative stuff, then all the, you know, the red tape and all the logistical side of things, that was done a lot quicker because there was a global collaboration and a global effort to really get us out of the pandemic. So, that was the speedy bit of the vaccine development, the actual testing and the actual manufacturing, all that kind of stuff, it wasn't rushed at all. So, don't be worried about why it was done in 6 months. The actual, you know, the real meaty part of it wasn't rushed, all the red tape was cut so we can get the vaccine now and protect people and get us out of the pandemic. So that's the key thing to take away."

Conclusion:
Everything was rushed, from unblinded trials, to irrelevant trials that didn't prove safety and real results. They knew exactly what will happen to the body when the mRNA will enter the body. Mr Karan should ponder what he's done and said here, online and to his patients, because he will eventually have to answer to the people he's hurt. He obviously did not read any of the official documents of the trials or of the jabs. This is unacceptable for a medical professional.

Question no 15 (asked by the authors of this book):
How will you create immunity against the real virus when the vaccine does not contain any parts of the real virus?
Mr Amir Khan (NHS GP) responds:
"Um, so you're absolutely right, the vaccine does not contain the whole of the real virus, but what it does contain is information that tells you about parts of the virus, and one of the most important parts of this virus is, we've all mentioned at some point, it's the spike protein. So, you may as well have seen images of the virus on the news and stuff, it's like a ball with little things sticking out, and those little things sticking out are called spike proteins and when the virus, if you breathe it in or gets into you, the spike protein binds to one of your cell, fuses with it and

allows the virus to get inside. Now, we only need our immune system to recognise parts of the virus to attack it and kill it, so these vaccines get your immune system to recognise just that spike protein, part of the virus, so when you are, when you do come across the real coronavirus, the idea is that your immune system has been primed already by these vaccines and it will recognise the virus immediately and it will, it will call its army of white blood cells and things that will attack the virus, and pull it apart and destroy it, before you even know you've got it, before you even get sick from it, so that's the idea. You only need to recognise part of it. It's a bit like, um, if you're in a supermarket. I don't know about you, but when I was in the supermarket, and I was, and I said to my mum I'll go and look at the toys and then I'm looking for my mum later, after I've found the toy I want, and I could just recognise her by the shirt she's wearing, or something like that, because I know in my brain what part of my mum I will recognise immediately. It's a bit like that, that's what the immune system does, it recognises part of the virus and it will attack it before it makes you sick."

Conclusion:

Another answer from Mr Khan proving that he does not know anything about immunology, viruses and disease. First of all, you need a high viral load in order to get sick with a coronavirus. One cell will never be enough to make you sick. It can take up to thousands of viable viral cells to be able to overwhelm the first barriers of which the macrophages (modified white cells gobbling up discard from the body and turning it into nutrients) are a big part. And when you do start to create antibodies, they will not start rotating the virus to see where the spike protein is. Antibodies need to have the right chemical composition to penetrate the virus cell from any angle. You can only produce the right ones if your white blood cells will synthesise the real virus, all of it. If you do develop antibodies against the spike protein it is irrelevant for staying healthy. When you do get infected, you will have no protection, as these antibodies will not be able to attack the virus. They are not fit for purpose and the virus will find it easy to take control. On top of that, because the spike protein produced as a consequence of the mRNA instructions delivered to your cells is latching itself to the membrane of cells and creating clusters or clots, the immune system will be overwhelmed by what it needs to clean, making you more vulnerable to serious disease from any virus, not just COVID-19. Mr Khan is a stranger to medical science.

Question no 16 (asked by the authors of this book):
Why should I trust the companies making the vaccine when they have poor ethical and legal histories?
Ms Faith Uwadiae (immunologist) responds:
"Um, in terms of trusting the companies it's very similar to what everyone said here already. All these,

um, drugs, have been through vigorous testing, they've been through all clinical trials, phase one, phase two, and phase three clinical trials, and um, basically being approved by all our governmental bodies. They're completely safe to use, so even if you felt in the past that there was some legal or ethical issues that you may have not trusted, know that in the current days they have been through proper vigorous testing and approval from real legal bodies. So, yeah… Take them with that off it."

Conclusion:
In 2023 we know as a fact that what Ms Uwadiae said in 2021 is completely flawed. Corruption is everywhere, especially in the government, including the MHRA and the FDA in the US. These jabs never passed safety protocols in any of the phases, and phase three was the real world, with booster after booster after booster. And now, there is an 84% increase in the mortality rate of millennials, while the mRNA jab manufactures are being protected from legal action.

Question no 17 (asked by Uber or another driver):
I have a specific health condition, so blood clots and high cholesterol. How safe is the vaccine for me? Is it gonna put me at a higher risk?
Mr Amir Khan (NHS GP) responds:
"Ok, so if you've got an underlying health condition, you're at higher risk of developing complications from getting coronavirus itself. So, you should be

prioritised for the vaccine because it will offer you really good protection. Um, anyone with any underlying health conditions really is at increased risk more so than those without underlying health conditions from developing serious illness from the coronavirus. With regards to blood clots, it's, it's, I can understand it is confusing. When it comes to this possible association between blood clots and the vaccines, it's very specific types of blood clots that we're thinking about. It's, it's, it's blood clots that are associated with something called low platelets which are components in your, in your blood, and the vast majority of people who've had blood clots in the past have not got this type of blood clot, um, so, if your blood clot is related to something that happened in your leg perhaps or in your lung, chances are it isn't this specific type of blood clot, and, you know, if you've got high cholesterol it could be that you developed a clot because of risk factors like that, um, then, then, you needn't worry about the blood clots associated with the vaccines, and its really, really important that you go forward and get the vaccine because it is far more likely you're gonna get a blood clot from getting coronavirus than from the vaccine."

Conclusion:

What Mr Khan said above is probably the most dangerous statement so far of the whole interview. Blood clots are dangerous no matter their type and the reason that caused them. When you have blood clots you should panic, because if they travel into the lung or brain, your life is in grave danger. We all have small blood clots at some point in life due to how we live and what we eat, but these are types that get disintegrated naturally by the body's defence mechanisms. But some blood clots that are bigger in size pose serious dangers. A coronavirus will almost never give you blood clots as it is a respiratory virus, not a virus that can pass through these barriers and attack other internal organs. There will be an accumulation of white blood cells at the place of infection (inflammation in the lungs and throat mucus lining), which could create small blood clots, but they will be "melted" down as you start to heal. Those will not be profound blood clots to endanger your life. If you already have blood clots in your body due to a certain medical condition, taking an injection that will give you more blood clots is a clear danger to life, as your body will have exponentially more issues to fight off. And being put on anticoagulants is not a long term solution, as they are thinning your blood too much. When it comes to the platelets, the ones that had the mRNA jabs are experiencing a very low count because the spike protein that their bodies are producing attaches itself to all cells as it travels with the blood stream. It'll destroy their membrane and put them together in bunches, whether they're white blood cells, red blood cells, platelets, etc. Here lies the extreme danger Mr Khan is refusing to acknowledge. He's not aware of this or he knows nothing about vaccine therapy altogether. In any way, he is, as one might say, a damn right criminal professional with no moral integrity and specific medical training.

Question no 18 (asked by Uber or another driver):
How safe is the Astra-Zeneca vaccine, especially when you're under 35?
Mr Amir Khan (NHS GP) responds:
"Yeah, should I go again? Only because I mentioned it a little bit before. Um, if you're under 35 you won't be offered the Astra-Zeneca vaccine for the vast majority of people, um, because we, in the UK we are offering anyone under the age of 40 the Pfizer of Moderna vaccines, and we've secured enough doses of them to make sure that is the case. Um, so, it's unlikely. The only reason you may have had it, if you already had it, is if, a, you were called for because of an underlying condition before those age guidelines came into place, um, in which case, if you've got an underlying health condition you're at increased risk of developing complications from getting coronavirus, so the benefits of getting the Astra-Zeneca vaccine outweigh the risk. But the vast majority of people under the age of 40 who have no underlying health conditions will be offered the Pfizer or the Moderna vaccine."

Conclusion:
Once more, Mr Khan is just playing the guidelines script, guidelines that have no scientific basis whatsoever. He keeps on repeating that the benefits of

getting the vaccine, any of them if you don't have access to one that's "right" for your age group, outweigh the risks posed by the real virus. This has no scientific support whatsoever, as it depends on so many factors that your case can never be properly assessed by any doctor to recommend getting an mRNA injection. Mr Khan should know the risks of this therapy, known way before this therapy started being injected into people. The specialty literature is quite thick about the implications of reprogramming the human cell to produce alien proteins. Most times it is called CANCER! Mr Khan does not know anything about human medical conditions and the mRNA technology. He was paid to promote this therapy to Uber drivers.

Question no 19 (asked by Uber or another driver):
How do we boost our immune system? What's your advice if you're not feeling well after the vaccine?
Mr Karan Rangarajan (NHS surgeon) responds:
"Yeah, in um, you know, breaking that down into two parts, um, you know, in terms of boosting your immune system, a, you know, there are various things that you can do just in your, you know, daily routine ensuring that you get adequate sleep, you know, having a regular sleep schedule, um, sleep is a time when your body repairs itself and, um, low, you know, sleep deprivation and poor sleep is associated with increased risk of infections and a poor immune system, so making sure you get enough sleep?

Another thing is making sure you get, um, regular exercise and it doesn't mean any heavy weight lifting or sprint, just, you know, going out for gentle walks, um, you know, 20-30 minutes a day, um, exposing yourself, um, you know, to that kind of, um, cardio vascular, you know, exercise or anything similar will boost your immune system because exercise has been shown to, you know, strengthen your immune system as well and make you, um, you know, protect yourself against COVID, um, and, you know, in terms of, um, sorry, what was the second part again? Yeah, and, you know, as I mentioned in one of the questions previously, if you are feeling unwell after getting the vaccine, some mild side effect are to be expected, a headache or feeling, um generalised body aches, and, you know, just feeling like, you wanna, you know, spend the rest of the day in bed, it's completely normal, um, and it shouldn't last longer than a few hours or a few days at most. But if you're having persistent problems and having concerns, um, you know, with regards to serious, um, health concerns that you're having or you are worried, um, you know, in the UK you can always call 111, or you can, you know, get in touch with your healthcare provider, um, who can direct you if you're having any specific problems but the vaccines are very safe, and like we've mentioned before any serious side effect are incredibly rare, one in several millions so, you know, it's acceptable to have some muscle aches and mild symptoms, which will go away, you know, within a few days."

Conclusion:

The level of malpractice in the replies of these supposed medical professionals is terrifying. As you've probably noticed, Mr Karan does not mention anything about vitamins and good diet being at the core of the immune system, or even better, sunlight directly on the clean skin. The NHS does not allow any natural remedies to be promoted by doctors as solutions to disease prevention and management. Stay away from such doctors like you keep away from the plague. Combine that with zero belief in God and you've got a licensed criminal, as one might say. Protect your health from these people with the price of your life. What they will do to you is manage diseases that are killing you slowly. We have seen serious side effects from all mRNA jabs in a proportion of 100% of cases, as confirmed by nurses that administered the jabs in various centres across London. Drivers that were healthy after contracting the new virus, are now dying after getting the mRNA jabs or are seriously ill, with the mildest side effect not being able to recover from a simple cold (constantly snotting and with flu-like symptoms). The effects are not rare and the "safety" of mRNA jabs is non-existent. These are not fully approved, nor fully licensed, and they will never be. They're telling you that they are because you don't know what EUA means nor what pharmacovigilance is. The only way to protect yourself and your family from making a mistake based on their advice is to study medicine.

You don't need to know all of it, but enough to stay protected from malicious entities and "professionals" like these panelists. Learn to know where to look and what to search for, especially how to ask questions and how to get the right and complete answers. If no clear answer can be given to simple questions, detach yourself and watch everything unfold from the backstage. There is always a financial benefit for the oppressors in situations as this. This is why you will never know the full story. You can just read the patterns and connect the dots.

Question no 20 (asked by Uber or another driver):
What is the point of taking the vaccine if I still have to take the test to show negative when travelling abroad?
Mr Amir Khan (NHS GP) responds:
"Yeah, I'll go for that. So, Karan actually answered this earlier in that the vaccines, their main job is to keep you from getting sick, from, from coronavirus, um, and, and so, if you get the vaccine the main point of it, you know is to stop you from getting sick ending up in hospital, and possibly worse. So, that is the main job, of, of the vaccine. The reason you have to take a test, um, um when you come back from abroad, and remember a lot of countries are still on a –do not travel, um, kind of list– an amber and a red list, so be mindful of that when you're thinking about booking. Um, the reason you have to take a test is because we don't know how much they reduce the spread of the virus. There is some evidence that they do but it's not

conclusive yet, um, so you may well be harbouring the virus and may not exhibit any symptoms, a, a, an you could pass it on to people who haven't had the vaccine, a, and that could be older people in the countries that you're travelling to who haven't been vaccinated, but here in the UK could be younger person, and young people may not get sick from the virus and end up in hospital, but there's a real risk of this condition known as –long COVID– in young people, um, where they can develop symptoms that go on beyond three months, a, such as pain, shortness of breath, fatigue, the list goes on and on, and trust me, I see patients with long COVID all the time in my surgery, and you do not want it. Um, so please, please, please do go forward for the vaccine, and if you are travelling it is important to isolate and test, um, on return according to the guidance for that country you've been to."

Conclusion:

Mr Khan falls under the category of the typical GP most British people are complaining about. He knows absolutely nothing else but only what is being told by central authorities, in this case the NHS, an institution counting on medical advice from non-scientist magnates as Bill Gates and Fauci (who made fortunes during the last three years). First of all, you cannot be infected with a respiratory virus and show no symptoms. Even when you are feeling sick, it takes a very high load of viral cells to get someone infected, plus other conditions that are required to infect. This is old science that Mr Khan does not know or did not want to mention. But apparently he knows everything about vaccines (the only viable solution), which he probably never studied or appropriated. People like him don't know essential medicine. For example, during the first two years of the pandemic, after contracting the new virus in February 2020, we have not got any of our riders infected with any virus, as we've had plenty of recurring riders in the vehicle. They didn't infect us with anything either for all those years. Getting infected with the same virus (even possible new variants) twice in the span of one or two consecutive years is almost impossible if you are not heavily immunocompromised. This is the real science we have worked with within Bio Products Laboratory in the UK, the former Lister Institute. You can carry the coronavirus with you, but you can only infect others if you are infected and showing symptoms. That means that you are carrying a high active load. If you are carrying a small load (no symptoms) it means that you're immune already, and that you cannot produce a dangerous load to spread the virus to others. You cannot produce new variants either. They played the asymptomatic card because their plans wouldn't have worked otherwise. So, only vaccines can be the way out of the pandemic, masks work (they don't – we know the PPE procedures), self-distancing works (it can never work and it'll destroy human interaction – it'll desensitise the population, increasing selfishness), etc. There is no *long COVID* either. Long COVID is made up of patients that haven't received proper medical attention when having a high temperature, or doctors misdiagnosing other conditions as *long COVID*, like hypothyroidism for example (it makes you feel tired, without energy, heart problems, etc.). A vaccine will never stop transmission. No matter the vaccine taken, you'll eventually still get the new virus too. Vaccines are meant to ameliorate the symptoms when you get it, so you don't get a bad episode. But a bad episode of coronavirus is almost never lethal, whether it's a new virus or old. Your immune system knows what to do. It does matter to get proper medical attention from doctors with high morals when you have a high temperature. Mr Khan does not fall in that category. We have seen people with high temperature who weren't given any medication and they stayed like that for two weeks or more. If you have a high temperature for more than a day, your life is in grave danger, as you need immediate medical attention (antibiotics – it means you're having a secondary infection). Testing for a coronavirus is just another way for the corporations to squeeze more money from you. They would've been free if they worked, but any test cannot give you the virus or if you're infectious without considering symptoms, when they first appeared and RNA sequencing (which can take even a week), because from the fourth day of symptoms onset you can no longer infect others if you're a healthy individual. Now you know that you need to study virology in order to protect your health. We recommend general medicine. You must learn medicine to become immune to untrustworthy, shady or bogus individuals such as Mr Khan.

Question no 21 (asked by the authors of this book):
The mRNA technology was considered highly dangerous pre-2020, as Moderna gave up the research with mRNA due to long-term negative effects on humans? Why was this tech not used before if RNA vaccines are less effective?
Mr Karan Rangarajan (NHS surgeon) responds:
"Yeah, I mean, you know, there's, you know, these, these companies, these manufacturers which make vaccines they don't just make vaccines, we know that, you know, test a lot of other things as well, there's dozens and dozens of trials going on at the same time as they, you know, they're doing other things as well, and sometimes at one of the trials, for whatever specific thing they look at, for whatever reason the data doesn't support it and they move on to another thing, and they tweak things as well, it's just like when you're, you know, you know, sort of, um, building something, there are, you know, things which you tweak along the way, and, you know, just because they stopped the trial because, you know, it didn't go as expected, it doesn't mean that it's a harmful effect on something. It's something which, um, you know, they might go back to or something they, important they learned from that trial, which they'll then use for future trials, so, you know, the key thing going forward is that, you know, the mRNA technology wouldn't be used and wouldn't be made widely

accessible for, you know, a global population if there were significant safety concerns. And, you know, we've had data now on pregnant women getting the vaccine, mRNA vaccine that is, a, you know, and that it's gonna be approved for young children as well soon, adolescents between 12-16, um, you know, we've got elderly people, people have got other medical conditions, all these different, you know, people from different backgrounds, different ethnicities, different, you know, populations all across the world have received the mRNA vaccine without any major safety concerns, and, you know, that has been backed not only with real-world data but trial data as well, and we've got now long term data, because the first people have received their first mRNA vaccine back in December 2020. We're now obviously towards the end of May, almost at June, so, you know, we've got almost seven months of long term data on people, and and that's not including, a, longer term term data, almost a year of longer term data from the people from the trials who got the mRNA vaccine and they've not had any problems almost a year into having their vaccines, so we know it's very safe and the reason mRNA technology is very fascinating and very good is because we can rapidly change the strain we use in the vaccine to target new strains. So, if there's another variant that, you know, came within the next few months, we can, and that was shown that the new variant, you know, hid away from the vaccine, or evaded the vaccine, we can quickly change the vaccine to combat the new strain with mRNA technology and we can change the vaccine within a couple of weeks they say as well, so we know it's very safe and very effective."

Conclusion:

There you have it, Mr Karan, a surgeon in the NHS. Absolutely nothing changed in the original mRNA technology and the risks from when it was modernised back in 1986 by Dr Robert Malone (shamed for speaking against his discovery being given to humans). Absolutely every scientific paper on mRNA trials is horrifying to read, and all studies that show safety and no significant side effects have massive flaws in the data used and have been highly manipulated and fabricated to fit the narrative. That manipulation of data involves a lot of hiding of contradictory safety results. Feel free to read all these studies. We guarantee that you'll be shocked, whether you've got medical knowledge or not. Mr Karan is saying that in medicine there is a silver bullet for all. That is actually the biggest mistake a doctor can make. We truly believe, besides the fact that he's been paid by Uber with our money to promote the mRNA injections to us, that Mr Karan might be part of a secret society. Only such a person can be so at ease with damaging human health. Your cells should never be forced to produce an alien protein, and that is what the mRNA technology is doing to your body. What follows is your immune system trying to take out these cells, which is called an autoimmune disease,

AIDS (as it happened with the Hepatitis B shots in the early eighties when they blamed it on HIV –to which many humans are naturally immune–). There is long term data on mRNA vaccines from way before 2020, and it is against any current narrative. They couldn't use the real virus to make RNA vaccines because COVID-19 is a patented virus. Plus, the mRNA technology is more costly compared to RNA technology. You have been lied to all along. They've poured billions of pounds into building the infrastructure and logistics to scale it up, not to develop the mRNA technology. And as you can see for yourself, Mr Karan is making no sense whatsoever when answering, proof that what he knows about this technology is what he's being told by the NHS. Starting with 2020, the National Health Service turned into No Health Service… And yes, Moderna did give up mRNA research for a while, or at least they pretended, motivating high risks for animals and humans. Their fears confirmed in 2022.

Question no 22 (asked by Uber or another driver):
I had COVID-19 in January but now I'm ok. What is the reason for which I need to take the vaccine?
Ms Faith Uwadiae (immunologist) responds:
"Ok, so, um, the COVID-19, if you had COVID-19, you would've developed some sort of immune response to it, so you'll have antibodies and you'll have cells in place to try and protect you from disease. But the thing with natural infection and

going through disease is you can't guarantee the level of antibody response you have or the level of protection that you have or how long that protection will last necessarily. So, the one good thing, and the reason we still advise people to get the vaccine even if they've had COVID-19 is that you essentially would get a strong response with the virus, um sorry, with the vaccine, that is guaranteed. So, you have a guaranteed level of protection that you wouldn't have necessarily with the natural infection, because depending on the level of dose that you were exposed to of the virus, um, when you initially had the COVID-19, that would determine the level of immunity that you have, but with the vaccine we're giving you a strong response with dose one, and then we're boosting that further with a second, um, antibody, sorry, with the second dose of the vaccine. So, the key thing is it gives you a good level of protection, you never know, even because you've just had the vaccine, sorry, just cause you've had COVID-19 you might have an even bigger boost when you have, um, the, um, vaccine so you might find it even more beneficial. So, yeah…"

Conclusion:
Ms Faith is the first immunologist I've listened to that doesn't know immunology. Little did she and the others know that on the other side there were people that have seen and worked with tonnes of human plasma and antibodies outside of the human body. Let's get something straight. When you get sick and

then recover from a respiratory disease caused by a virus, you are then immune to that specific virus. Otherwise you'd die. Of course, the levels of antibodies between individuals and sickness episodes cannot be the same. That doesn't mean you're less immune to the virus. Immunity can never be boosted! You can only live a healthy life to support your immune system. You never boost the immune system. Your body works in a holistic way, with signals and substances being produced out of the ether. By having the disease and then a vaccine with additional harmful contents, you'll confuse your immune system and force it to spare no resources. Vaccinations overwhelm the immune system, and this is a system you do not want to stress on a regular basis. It is what keeps you alive. When your immune system gets destroyed, you die, so you need to be really careful with how you're treating it. You do not need a vaccine, especially if you've had the virus first, because you're immune for life due to dendritic cells (dormant memory cells). You will get the virus again, vaccine or no vaccine, but it'll get easier. Plus, an infection occurs slowly, whereas a vaccine is delivering a huge load at once, creating panic in the body. If the real virus would infect you in the same way, you'd have almost zero chances of survival, no matter how healthy you are. This is why you have barriers. The tonsils are first to produce white cells in the throat triggering the immune signals, then if they fail, the virus goes in the throat and synovial mucus (filled with antibodies), then if this fails (you got a big viral load), the virus passes into the lungs, where many white cells (dendritic cells capture the virus for analysis) are ready to create inflammation triggered by the previous barriers. And the infection stops there. Then IgM and IgG is being produced when the big clean is under way. By the time your lungs get infected, you're already in recovery mode, as that can happen after 3-4 days, when you start to feel your lungs heavy. It has been proven scientifically a long time ago that you are no longer infectious after about four days from first symptoms. Ms Faith, despite her karmic name, resembles a qualified criminal for promoting a therapy she knows nothing about. She should be ashamed for not knowing if you're immune after having COVID-19 and recovering just fine.

This was the end of the presentation sponsored and paid for by Uber. If you want to know more about the effects of mRNA/mDNA therapies look at what Zolgensma is doing, a medical company producing therapies for children born with SMA (Skeletal Muscular Atrophy). The same life-ending symptoms are identical with the ones following mRNA vaccination. Another unapproved medicine they used to one might say kill patients in hospitals is Remdesivir (do your own research). Plus, the Zolgensma shot costs a whopping £1.7 million. Some might say, "yes but its only one shot and it's normal to cost that much compared to producing it at a large scale". No, that is not true, it's just a rip off because it is paid for by insurers in the US and the NHS in the

UK. Why not experiment on public money? If doing a PCR test costs so little, then there is no reason to charge almost £2 million for an mDNA shot. To explain all these processes it'd take another book, but it should cost a couple of thousands at most. We have the technologies and facilities to produce medicine for a residual cost, no matter the scale of the operations. Zolgensma is experimenting on children, with many of them dying post-therapy, caused by the mDNA shot.

Ok, but what should Uber have done? They should've been impartial and not pay untrustworthy professionals to promote a therapy they know nothing about. Uber is not a medical authority, and they presented no RISKS! No Uber employee is a doctor in medicine, immunology, or vaccinology. For this, Uber has harmed all drivers that followed their advice and took the jabs who are now in peril. There will come a time, sooner than later, when Uber will have to pay hundreds of millions of pounds, if not billions, to their drivers no matter if they had the injections or not. They've done it with a clear intention to harm, as none of the panelists gave any information about how these jabs work and if there is any associated risk. If the presentation would've been about investing in FOREX, all participants would've debated only the risks and none of the get-rich-quick benefits. The day of reckoning is getting closer for Uber and their guests from the presentation. At present, we can smell the vaccinated with mRNA, even through strong perfume. They all have an unmistakeable identical rotten smell, similar to the bedding smell in cheap hotels. It's difficult to describe it but we're sure you can feel it too if you haven't been jabbed. You're not crazy, it's true. This is a bigger medical tragedy than the thalidomide disaster. This confirms old science about the mRNA therapies, that they're modifying the chemistry of your body, for life. Knowing what we know in 2023, what Uber has done is shameful, unforgettable, and unforgivable (some could say criminal)!

What should the doctors do? They must practice medicine no matter what. The NHS is there to manage their patients for them, but never to tell them what to do or how to practice. If the NHS is telling you what to do as a trained doctor, why did you go to medical school in the first place? It means that anyone can be a doctor, even us. This is why malpractice is one of the main reasons for death in the world, because doctors are listening to their employer and Big Pharma, instead of practicing the medicine they learned in medical school. Uber's guests, during this Q&A, have behaved shamefully and disgustingly, and acted against their oath, betraying the medical science. Even if you own a PhD in medical science but not in vaccinology or immunology, you're not allowed to promote something you know nothing about, just because you're being told to do so. As a doctor you are never being told what to do. You do what you know with certainty and confidence. If people with no medical knowledge (superstars/ famous individuals and influencers) and medical professionals that haven't been trained in immunology were allowed to promote these jabs as being safe, why were the ones with the same type of credentials that spoke against them being blocked from speaking in public? Aren't the two groups supposed to be as dangerous?

Categories of people that have told us to stay away from these jabs and any other future ones: British MPs, rich individuals across London, doctors with decades of experience, nurses, freemasons, businessmen, bankers, religious figures, Jewish people from Golders Green area & Rabbis, very rich individuals and politicians from Arab countries like Saudi Arabia, individuals with family members in secret societies and politics, etc. This, cumulated with our own medical experience and knowledge about the world and life, was enough to make us stay away. You do not always need to know everything, but just enough to protect yourself and your family. You must never forget what happened in the past. If all humans would have studied about what happened during the 1918 "Spanish flu" epidemic (most that died were vaccinated), nobody would've trusted any government, doctor, or pharmaceutical company in 2020.

44. Could Uber Stock Plunge To Under $1 Soon?

All fundamental and technical metrics might indicate so. Despite Uber looking good on the stock market, what is happening in the background with the company does not match the painted image. This is what a savvy investor should have access to, before looking at the technicals. Not knowing these details about the true state of the affairs of a company that does really bad, can lead to huge loses for small and big investors. Uber's stock price does not reflect the health of the company. Being a listed corporation, its stock price is being manipulated by Wall Street mechanisms and lies, while big investors are pumping and dumping the stock every day. When these big *whales* will find out about the many billions of pounds in revenue that will not flow into the company from trips anymore, due to drivers quitting and riders not using Uber services anymore, there will be mostly dumping the stock in great quantities by former Uber backers, while questions about the elephant in the room will be raised within the financial world.

In short term, Uber's situation looks promising, but those good times are gone now, not because of the pandemic, but due to how it's been run since 2020. This is not happening only in the UK, but around the world where Uber is present. The "green" agenda being pushed forcibly onto drivers will be Uber's downfall in the next few years, and it will be a very high and painful fall for the company. Uber thinks that it's doing the right thing by not looking after its partners. They do not understand that their drivers are humans and independent, with a powerful will to work ONLY if good and decent money is involved. The switch to fully autonomous vehicles will be even more painful for Uber and will probably lead to the total collapse of the platform. There are transportation companies that are way better prepared for these agendas of the future, and one of them being Tesla, a monopoly waiting to happen, and competition Uber will never be able to compete with once they start releasing their autonomous ride share service. The storm is coming from all directions for Uber, and its inevitable crash will happen before being prepared for the future they're participating in. Today, Uber does not have money for the technological advancements they promised. Uber's CEO announced huge profits after the pandemic, but that will be short lived, as the big investors will pull out sooner than later, while the bank is full. This is what they do when they smell incoming headwinds. The company will always paint a favourable picture to avoid this from happening. It is unavoidable though, as drivers' patience cannot be stretched indefinitely.

The breaking point for driver exodus is getting closer, as they cannot drive without making money plus a profit forever. It is barely survival for all drivers on the Uber platform, regardless of how busy it is. Usually, there is an approximate increase in demand of about 30% post-pandemic, and drivers are earning 50%-60% less compared to pre-pandemic levels. That is a tragedy, and it is bound to cause a rupture. Many drivers already had other businesses they were pumping profits from working with Uber in, but that stream is now over. These drivers will give up doing it. Despite having plenty of drivers on the roads, Uber still needs to activate thousands of new drivers each year. Word has gone around about the struggle on the platform to earn survival money, and no one wants to become a partner driver anymore, with any platform, not just with Uber. Hence the number of Uber drivers quitting is way greater than the number of new drivers. Where will the money come from for Uber? Another problem is all the restrictions put on private hire drivers when it comes to driving in city centres and switching to electric vehicles. Many who've driven electric vehicles on the platform realised quickly that it is financial suicide, having twice the costs of driving a hybrid-petrol vehicle. And so, these electric vehicles are being returned to dealers, while rental companies are struggling to rent them back to other drivers. This is an ugly, vicious circle, a nightmare for drivers, rental companies with any type of vehicles (but especially electric), and riders.

And then we have the riders, who are switching to other transportation alternatives that are readily available, like taxis, no matter the price of a ride. More and more former riders are saying that they stopped using Uber for all the reasons explained in this book. In their vast majority, they know that Uber as a company and business model is finished. It is also a known fact that bringing in autonomous vehicles to replace its drivers is not viable due to safety concerns, regulators not allowing them to do it, and the government not wanting people to move so much. The future they're promising us will not happen any time soon, if global population does not drop significantly. Not being allowed to bring in autonomous vehicles is not being considered by many investors when assessing Uber as a possible good investment opportunity, which Uber stopped being a long time ago. There will be a recovery of the stock market, but Uber will be the biggest loser of any recovery when all this information will reach the investors. And in case of a market recovery, if Uber will not succeed in capitalising on it even at the peak of the surge, if it will fail to keep hold of its big investors during times of market hype, we believe that its stock can actually go way lower than $1 at the next crash. Today there are higher chances for consecutive crashes than positive market sentiments. As we have all witnessed so far, the new capitalists are working tirelessly to destroy all industries, for the "good" of the planet. Impressive surges will be a thing of the past, and when they will happen, they will be short lived. Another aspect keeping new investors away is that Uber does not pay dividends to its shareholders. When existing ones will pull out, it will be a disaster for the company, and for whoever is left holding the bag. Uber will most definitely not be one of the best performing players on the stock market in future uptrends, if they will still be listed altogether.

      Another alternative for Uber to fight off or cover up the coming headwinds, would be to take the company private through a buyout of the biggest investors. But this would truly mean financial collapse for the company, involving a huge number of lawsuits around the world. We do not really think that this is an alternative for Uber, but whatever their outlook on the long run would be, it cannot be a positive one. Their empire in the UK is crumbling faster than they can pick up the pieces, and a good part of this situation is intentional through corrupt and colluding management. They had to continuously compromise with the discriminating authorities at the drivers' and riders' expense, and this will cost them big, globally. Whoever that executive that slept with the MP's girlfriend/wife was, he's done a good job in crippling the Uber corporation. And if it's ever proven that an MP waged war on Uber from a personal feud that's affecting millions across the UK (as per Canary Wharf bankers rumours), a full criminal investigation should be initiated immediately. Never forget that the push against Uber only started suddenly and for the wrong reasons, which at least raises suspicions with regards to political involvement.

The Action

The following is a document sent by the authors to the Employment Tribunal of the UK Supreme Court for the rescinding of the decision to uphold the High Court's ruling to treat all private hire drivers on the Uber platform as workers. This document is a model specific to the situation of the authors, minus the personal details. You have to shape it around your personal situation/circumstances and send it individually, not as a group. DO NOT JOIN UNIONS BECAUSE YOU ARE AN INDEPENDENT CONTRACTOR/SELF-EMPLOYED, NOT A WORKER! The more drivers complain to the Court the right way, the higher the chances that things will change for the better. If you have more relevant details add them to the document before sending it to the Court. Each document sent to them is a unique case, and their aim is to uphold human rights and protect individual rights. In order to avoid the rewriting of the whole document, which is quite lengthy, use a device to photograph the pages containing it, and then copy and paste the text from the photos. Most new devices allow text to be copied from images. You just need to make sure the page is flat when photographing it, so the copying is precise afterwards. If you have any additional questions don't hesitate to reach out to the authors via the channels provided on the copyright page. Send this document by normal post as priority with confirmation of delivery. You will notice the use of "Respondents" and "Plaintiff" interchangeably because, depending on the state of the legal proceeding we are referring to, they have different names. These words are being used in law to determine the party/parties in the legal proceedings. "Aslam and others" were "Respondents" during the last stage of the lawsuit with the High Court, but they formed the "Plaintiff" when Uber appealed with the Supreme Court. You must pay attention to the modifications you are making to the sample document, otherwise you risk looking unprofessional in front of the Supreme Court. If you're not sure, do a bit of research, or ask someone with experience. We are here to help!

You also need to contact Transport for London too, in order to get the licensing conditions changed back to allow the contract to exist between the rider and the driver. Also, to stop Uber from charging drivers' riders more than 25% of the fare. There is a working business model and that is the old model. Drivers should not earn less than 75% of the total cost of the trip, including tips. Platforms as Uber are just intermediaries. The driver is offering the transportation service, not Uber. If all drivers went offline, Uber would have no business, despite it expanding into ticketing for public transport, taxi services, boats, helicopter services, etc. Send individual letters demanding the above from TfL. Shape the letter of complaint and request on the same lines as the following "Skeleton Arguments". If you get no reply from TfL, or a dismissive reply, challenge the court decision that forced Uber to change the contract after TfL's licensing legal action. Individual letters will turn the tide and give you a lot of power. You are a free individual and you need to claim this right. Individual interest is above any collective union action. Don't forget that it was a minority of drivers that changed your life on the platform, without even thinking of asking for your opinion, or that any changes should only be applied to them. They basically stomped on your freedom of choice. You need to understand that throughout the legal proceedings against Uber for worker rights, the courts have been misled by the founders of ADCU union. They never represented all drivers but they made it look like that without your permission and opinion. This letter is meant to tell the Supreme Court that one of their decisions has caused you harm.

Send your letter to:

The Supreme Court
Employment Appeals Tribunal
Parliament Square
London
SW1P 3BD

or a shorter version to each individual Judge that reasoned on the decision. Alternatively, you can ask for permission to appeal the decision, which costs between £800-£1,000. If all fail, you go to your MP and the Parliament.

Sign our petition here: **join our Discord server for details**.

Date:
[Insert Date Of Writing]

Skeleton Arguments for the rescinding or amending (to apply only to the Respondents) of the UK's Supreme Court Decision through its Employment Appeal Tribunal (EAT) in the final Appeal stage of the case Uber BV and others (Appellants) vs Aslam and others (Respondents)

JUDGEMENT GIVEN ON

19 February 2021

Heard on
21 and 22 July 2020

Author:
[Your Name]
[First Line Of Your Address]
[Town/City]
[Postcode]
Email: [Your Email Address]

I am respectfully asking the Supreme Court of the UK to rescind or amend (to be applied only to the Respondents) the Decision in the case Uber BV (Appellants) vs Aslam and others (Respondents) which settled that private hire drivers driving in partnership with ride hailing operators, in this case Uber, are workers.

As a self-employed private hire driver licensed by Transport for London, I have been driving with Uber as partner driver since [Enter Date], and have developed into one of the best drivers in Greater London area, achieving a service star rating on the platform of a constant **[Enter Your Rating]** out of a maximum of five stars. I have completed over **[Enter Number Of Trips]** trips to date without any traffic incidents and always with safety and comfort at heart. My activity on the Uber platform has been continuous to the present day.

Being a British citizen and living in the UK, my income is being spent actively in this country. Prior to deciding to become a full-time private hire driver I have worked as a **[Enter Previous Occupation]**. The decision to give up that career was easy to make, as full-time private hire drivers could earn **[Enter Number]** times more a month, up to March 2020. During the skeleton arguments I will compare the current situation of drivers with the market conditions before the UK's first pandemic lockdown.

The arguments this request is based upon are as follows:

1. **Aslam and others** have brought a case in front of the Supreme Court containing erroneous and bogus details. The subject of the case has been fabricated.
2. The **Plaintiff** argued that private hire drivers working in partnership with ride hailing platforms, did not get paid the national minimum wage, paid holiday, and pension. From the moment I have started working with Uber, there was not one single day of activity on the platform when I have earned less than the minimum wage at any time, even from the time I went online. This argument as basis for the legal action is totally flawed as I can prove it with my own activity on the Uber platform **[Only If You Started Working Before The Changes In March 2022]**.
3. **Aslam and others** have registered themselves as a union, have marketed as representing (all) private hire drivers in the UK throughout the proceedings, and by doing so misleading the Court. At the time, I was considered by law to be self-employed, and I have never given the Plaintiff my consent to represent me or my legal rights in such critical matters, nor did they contact me in any way to give me a chance to object. I was an independent contractor and no one had the right to disrupt my activity without prior consent. This was an unbecoming breach of my legal rights in a free market and a distorted representation of reality in front of the High Court.

4. The main parties representing the Plaintiff, Mr Yaseen Aslam, Mr James Farrar, and Robert Dawson were not active private hire drivers in the UK. They might have still had an active licence issued by Transport for London, but they publicly declared that they were not active drivers, earning income from unknown sources during the proceedings. They stated that they might return to driving on the Uber platform if they win. It is unknown if they did, and apparently at least two of them went back to previous activities. Mr Aslam owns an IT business and was recently granted a contract with the Ministry of Defence. Also, Forbes has presented him as a hero for managing to bring Uber down in Court and "to its knees". They did not mention once how many drivers agreed with this case by signing up to their union. I have personally spoken with numerous other private hire drivers and not one was part of such a union when approached.
5. Based on the Supreme Court's ruling, as I do not have a say in my statutory rights as integral part of my partnership with Uber, it goes without saying that Uber is legally bound to act accordingly and pay for my vehicle, insurance, licensing, medical insurance, etc. Uber cannot pick and choose, thus it must act as an employer. However, this would come with an even lower take home income for me, as I would make no sustainable profit, or any profit at all. Currently, all these benefits the Plaintiff has asked for are being deducted from the

fares through a higher percentage taken by Uber, which now ranges from 25% to 50% of the total fare paid by the riders, as opposed to 2019.

6. There are a few unions claiming to fight for private hire and taxi driver's rights, but most are not legitimate and have as only purpose to make money from membership fees while disrupting the working industries. By doing so, they are not just harming drivers, but millions of people depending on Uber to get from A to B as quickly, reliably and efficiently as possible, in London for instance. The union established by the people behind the legal action against Uber, ADCU, have always portrayed themselves as fighting for gig workers' rights, when gig worker means self-employed. They never represented me, I have never given them the power nor my consent to interfere with my business, because an Uber driver's business on the platform is not connected in any way with the business of another Uber driver. Drivers have the choice to be free by choosing their working hours and earn decent money, or be considered workers and earn survival money only – what is happening following the ruling and after Uber has changed its business model. There are other unions sharing the same views with ADCU, like GMB. I have tried to contact both unions to ask for the real number of private hire drivers registered with them, public information that should be freely available, especially to active drivers, regardless their membership status when affected by a Court

ruling, but I was denied an answer. In more recent developments, after the implementation of the Court's ruling in March 2022 by ride hailing platforms, and most drivers starting to notice and complain as to how the industry is being destroyed, once again these unions promised they will fix the issues with these platforms. There is a massive uproar among private hire drivers across the UK caused by the recent changes, when Uber and others had to implement the ruling. The media is completely silent when it comes to these matters, while Mr Aslam, Mr Farrar and Mr Dawson are making substantial money from the ADCU union membership fees. I have a vast experience with unions and all I have learned is that they always make things worse long term, with the most dangerous unions being the ones that form overnight, which are being run by people with no experience, getting backed by other questionable peculiar entities with a specific target and purpose. I have an extensive experience in these matters **[State Your Experience If Any]**. In this respect, I am of the view that the Respondents acted dishonestly, bending the law to suit their needs, as they did not have the right to raise such issues with the High Court on behalf of all private hire drivers, but only for them. As we see it, they presented fake evidence to the court in order to sway the verdict. They had the right to ask for changes applicable to themselves alone or leave the platform. What they were trying to fix has actually caused the rapid downfall of private hire drivers across the UK. Furthermore, ADCU are still showcasing on their website that they aim to represent all UK private hire drivers, which is not true, as well as biased. They do not and will never represent me, and I will never empower nor give my consent to any private entity to represent me in front of the UK's Courts. For this alone, I strongly believe that the Respondents should be trialled for contempt of Court for lying and presenting fake evidence to the High Court as well as The Supreme Court, and for distorting the truth during the legal proceedings. I do not know and was unable to find out from them how many private hire drivers signed up to their cause, but I can guarantee that they were just a small minority, as most drivers I spoke to were not part of any union. After significant research, I have discovered that ADCU, registered with Companies House as "United Private Hire Drivers Ltd", has been dissolved. Based on the same model, I could start a union today, even if I do not contract with the targeted corporation, make a bit of money from unknown sources, most likely from the union membership fees and "donations", start a lawsuit against the corporation demanding they apply my agenda with their contractors, and ruin the lives of millions across the UK. This paragraph alone should be enough to raise the alarm and beg the question as to what has happened, as the implications of destabilising the private hire industry are grave.

7. Being treated a worker has opened the Pandora's box for drivers and has given big corporations like Uber the liberty to act even more irresponsible and leveraged them to do as they please. Uber was bragging recently that their investors are very happy as their profits soared. They were soaring before these changes too, as they were giving considerable promotions to their partner drivers. Back in 2019, part of that was my money. Before being treated as worker, the issues with Uber were never about not making enough money, as in less than 50 hours a week of online time on the Uber platform, I was making somewhere in the vicinity of £5,000 per month. Today in over 60 hours of online time spent per week on the Uber platform, I am struggling to make a bit over £3,000 per month. I am only being paid holiday money every week, which is about 12.07% of what I have made the previous week. This amounts to between £60-£90/week. In 2019, I was earning ten times or more of this holiday pay just by working less and taking advantage of the offered opportunities. That is exactly what Uber was before March 2022, a great opportunity, whereas now it is just survival mode, breaking even and making less and less money on a weekly basis. The problems with Uber, before the lawsuit was filed, were as follows: rude and dangerous riders with low rating being allowed to use the platform without considerate safety measures for drivers, lack of complete transparency regarding accurate trip payments (I

had strong reasons to believe that Uber was withholding more than 25% and sometimes even from tips paid by riders), better compensation from riders for mess made in the vehicle, riders with low ratings paying more per trip, better compensation for being stuck in heavy traffic conditions, better compensation for late cancellations, getting detailed explanations from riders for rating the driver less than five stars and being fairer with drivers, fixing racial discrimination when organising parties/events (Uber were and are still organising big celebratory biased events, one might say discriminatory, as they were catered for Asian and Arab drivers only). Despite all the aforementioned, Uber still continues to be the standard in the industry. All functionalities provided by other active private hire operators in the UK and London alone do not entirely meet the industry standards in terms of safety and all of the above. There was no issue when it came to the overall amount of money I was making before Uber have changed their business model to comply with the Supreme Court's ruling. The money I am making now are less and less from March 2022, are highly likely capped to a certain amount, and the overall monthly earnings are not being kept in pace with the cost of living crisis. Being self-employed, I need to be able to earn as much money as it is available in the market at the specific time, place and availability similar to taxi drivers, which are not being limited as to how much they can earn at any given time or by any entity.

8. Uber implemented the Supreme Court ruling by changing its business model in March 2022. First, the contract is no longer between me and the rider, but between Uber and the rider. This means that I am being kept in the dark and consequently regarded as a "third wheel", despite being the only one having the power to start the contract between the two parties, execute it, as well as end it at all times when and if something goes wrong beyond my control (e.g. rudeness, violence, non compliance to Uber community guidelines, behaviour, mess in the vehicle made by riders, etc.). At the same time, I decide the level of service I am offering. Uber has no say. There are terms & conditions between me and Uber but the corporation cannot fully enforce them on me. I decide the level of service offered to riders contracting with Uber. There are plenty of drivers working with dirty vehicles and that is their choice. Because I am nowhere in the contract, despite having the power to affect the contract in a drastic way, Uber decided to stop showing me, the driver, how much the rider gets charged for the trip. With this, there is no transparency anymore. I am being kept in complete darkness while Uber can charge as much as they want, and then pay me as much or as less as they want. After thousands of trips under the current model, I found out that Uber is now withholding somewhere between 33%-52% from what riders are paying for trips. This translates into a severe decrease for the driver, from about a minimum of £1.25-£1.50 per mile to £1.09-£0.40 per mile in most instances. The effect of this practice is that more than 150,000,000 (one hundred and fifty million) trips are being rejected by drivers each month in London area alone, only on the Uber platform. This is my approximation based on how many drivers are registered with Uber in London. Many drivers had to give up driving on the platform altogether and are now in a poor situation after previously thriving when working with Uber. The driver cannot be kept in the dark about the charges, and I cannot understand as to why Transport for London agreed to issue a new licence to Uber with such a questionable business model. For the above reasons, I believe I am not a worker, otherwise I would not have the power over the contract between the two parties (Uber and its riders). Also, for safety reasons I cannot be taken away that power. If private hire drivers become employees, that will destroy the gig economy and chip away at people's freedoms and liberties to act as they wish and earn as much as it is available in the free market. The Supreme Court must uphold individual freedoms and liberties of each person, and repel actions that would hurt the majority. The ruling should at most apply to the Respondents.

9. If there were active drivers part of the Respondents, then they have either been paid to

join the action, have been grossly manipulated by the initiators into thinking that they could earn more with these benefits, or made to believe that benefits weigh more than making decent money overall without them. Whichever it is, they are wrong and form a small minority. The fact that unions refused to give answers to simple questions, that Transport for London is actively supporting this unproductive change from an outstanding business model that worked, and that the media is silent to increasing complaints from the drivers who bring a huge benefit to the UK economy, are all of these are red flags of a clear agenda behind this lawsuit aimed at crippling private hire drivers. Mr Aslam alone is now being portrayed as a hero by Forbes, despite the damages inflicted on this industry.

10. There is no secret that the current Mayor of London, Mr Sadiq Khan, was and still is a lobbyist against Uber drivers especially, who should be considered heroes of the pandemic. During 2020 there were no taxis or black cabs anywhere to be spotted in most of London, while Uber drivers were driving patients, nurses and doctors to hospitals, putting their and their families' lives in severe danger. I was one of those drivers as I never stopped working. I have driven thousands of nurses and doctors in the Greater London area, and many of them were complaining that they would've had no other means of transportation, including public

transport, to go to work or back home other than Uber. Many thanked me and called me a hero. I never dared to consider myself one, but in the current context it is in my detriment if I do not speak out about this injustice. I did not deserve to have my business destroyed on the platform, while others that could've helped were afraid of dying and they have now been helped to thrive even more. Despite all the facts, the ones that fell on their feet following this ruling are taxi drivers, Uber and others, TfL, and the main culprits, the unions misrepresenting and forging the facts in UK's Courts of law. There are no facilities to speak with the Mayor of London about this situation, nor with TfL (they do not reply to many email messages and there are no available phone lines), but it is truly unbecoming of the Mayor's Office to lobby for one side or another, and allow such business models to be implemented against private hire drivers. Double standards should not be permitted within a functional and unbiased legal system. For instance, even former black cab drivers are using Uber to go around, and they confirmed the industry's hate and subversive actions against any competition.

11. The choice to be considered workers or self-employed should lie in the hands of the private hire drivers alone, and ride hailing platforms should respect that. This implementation and statute change could not have come at a worse time too,

with the cost of living crisis affecting everyone and everything. I was expecting the demand on Uber to increase exponentially after the pandemic, and it happened, but as opposed to 2019, I am now earning around 50% (sometimes above 60%) less due to the implementation of this ruling. At present, I am in a very difficult financial situation not because of the pandemic, but due to the aforementioned legal factors, in the context of more business on the platform compared to 2019 (about 30% more), unrealistic transition to electric vehicles with no feasible financial aid, increased charges in London and airports for private hire drivers (despite Diesel hackney carriages still being exempt from paying them), exponential increases in living costs, etc. Also, I cannot afford to book time off and holidays anymore.

12. In 2019 I was making as much money as it was available in the market, in the area I was driving in. Today I am earning as much money as Uber wants. I strongly believe that Uber is limiting the weekly earnings by keeping more from the fares, often sending me many miles away to the pickup point for really low pay (e.g. distance of more than 10 miles with a real time duration of over 30 minutes for as low as £4.30), and by favouring other drivers that accept all trips (who actually earn less than the drivers who decline lots of trips due to them being underpaid). Regardless if I accept all jobs or not, I am making 50%-60% less than before the

pandemic in a recovering market with increased demand. Even before the radical change that took place in March 2022, paying 25% as service charge to Uber was a stretch, but I could compensate that with really good prices and working the right hours for me. In that context, Uber could afford to give me promotions that were many times attractive, as they were not sending me far for pickup and the prices were good, making the acceptance of almost all trips a good option to go by. From an acceptance rate on the platform of no less than 95%, now I am at a constant rate of 5%-10%. Most drivers have the acceptance rate within a similar range.

13. I am now being treated as worker by Uber following the Court's ruling, but as per the worker's law I cannot be a worker. I am still sending a self-assessment to HMRC myself. Uber is not withholding and can never withhold tax and national insurance in my name, because they cannot know my total running costs, as that would mean that I should give them in-depth private details about my health and the treatments I am undertaking, as well as details about my other ventures that fall under my general self-assessment. That would mean a total breach of my privacy and risks endangering my ventures by exchanging trade secrets with a second party that could use them in the market or against me.

14. GMB union are claiming that they are fighting for taxi and private hire drivers rights, but that is a contradiction as they cannot help one side without hurting the other. It is well known and documented that taxi drivers never liked private hire drivers despite all of us being people trying to make a living, just under two different existing laws. The tensions were always being created by the taxi drivers. In my first years in the industry I have been spat at, sweared at, showed obscene signs, all of them coming from black cab drivers. It took the cancellation of a few taxi licences for them to calm down and respect the community guidelines. Today that is not happening anymore, not to me at least. That being said, GMB are using false pretences just to make more money and revive long time forgotten issues. It is a known fact that the taxi industry have supported this lawsuit, they were quite outspoken about it as they knew what the effects of such a Court ruling would entail. What the Respondents have done, helped by the taxi industry or not, was a step too far and disrespectful toward the UK Courts of Law and me. From the moment I found out about the legal case, I knew what the consequences will bring, so I find it difficult to believe that these unions could not foresee the effects of it. Asking for minimum pay, paid holiday, and pension contributions, money I was making tenfold in 2019, is absurd and totally damaging. My intention is not for the Supreme Court to understand from this that I am name-

calling in a cause I know nothing about. This is a genuine concerning request coming from one of the best private hire drivers in London and its neighbouring counties, to let the Court know that the ruling in this case has damaged my life in a deep way, as it took away from me the freedom of making as much money as it is available in the market, whereas the Supreme Court's aim is to support and improve people's lives. Being treated as worker does not satisfy the full interpretation of the workers law. As I understood, this ruling applies to food delivery drivers too. In this instance their situation is probably much worse if Uber is keeping the same percentages from the fares paid by customers. They could barely make ends meet before the ruling, as it was.

15. The contract has to be between me as driver and the rider, so I can always know how much Uber is charging for the fare, because I am the one fulfilling the contract. I have make or break power over the contract between the rider and Uber, as the object of the contract, the service, is offered by me, the driver. Uber is just an intermediary platform that should not be in the services contract. I was trying to ask Uber for a full audit of my trips before the changes, but today this is almost impossible to obtain due to my legal status on the platform. Uber keep on saying that I am not allowed to know how much they charge riders, because I am a worker. As a driver, I never see the

original receipt the rider receives at the end of the trip. Uber never allowed that.

16. There are platforms such as Fiverr and Freelancer that act as intermediaries and contract facilitators between the service provider (in this case me) and the platform's user. People offering their services on those platforms are self-employed. They can never be workers. Ride hailing platforms are the same for transportation. This ruling has given free hand to all private hire operators in mistreating their partner drivers. Throughout my partnership with Uber, I have been referred to as "partner driver". Considering the highlighted circumstances, I cannot be a partner driver and worker at the same time, as one cancels the other, because being a partner driver means independent contractor, and I can never be an Uber employee nor worker per se if it was to interpret the implementation of the Court's ruling to its full extent. The whole ride sharing industry would collapse. I cannot be a worker because I can decide whether to work or not, when and where, complete the contract between Uber and riders, or cancel it altogether due to various reasons such as safety, breach of community guidelines, etc.

17. Treating private hire drivers as workers has many outcomes, and besides the unprecedented financial decline of drivers, there is a plethora of safety implications for riders too. Uber covers a more

extensive area than normal taxis and especially than black cabs do. In areas with unreliable or no public transport at all, booking an Uber used to be instant, as there is always a partner driver around. Before March 2022, drivers used to accept the trip requests in no time because the fares were most times sufficient for the mileage driven, especially during peak times, when the "surge" (exponentially more requests than available drivers) in trip requests was happening. At times, drivers could earn a whopping £60-£120 per hour. Under the new business model implemented in March 2022, the higher the price the rider has to pay to Uber, the more Uber will keep for its business, whilst the partner driver's pay per mile does not cover the actual terrestrial distance in real time traffic conditions, or is lowered so much that it becomes ludicrous. Up until now, it was not an issue if the rider would have taken me in an area not covered by Uber's service, because many times the "surge" price was enough to cover the miles driven empty handed until the active zone. Nowadays, that does not happen whatsoever. These practices are now putting lives in danger every day. A vast number of people depend on Uber, since it is now a service vital to the economy, as demonstrated previously, during 2020, when merely Uber drivers were taking patients, doctors and nurses to hospitals and back, with taxi drivers being afraid of the virus and who chose to stay home. It happens many times to pick up vulnerable people, disabled, women,

women with children, from dangerous areas where there are no other means of transportation available. Before the changes being implemented, they didn't find it difficult to book an Uber driver, whereas nowadays it can take a considerable amount of time. I have personally driven single women and mothers crying when picking them up, because they were scared of waiting long times in the area they were. This is due to Uber not paying drivers a fair price per mile and per minute, and the removal of the dynamic pricing which was accounting for real traffic conditions on roads at any given time. Even when stuck in traffic for long times, the payment to drivers does not change anymore, and when it's randomly adjusted, the trip gets frequently underpaid. Oftentimes, it's as if Uber wants drivers to spend 30-45 minutes on a trip in congested areas for a total pay of just under £5. Back in 2019, these issues were almost inexistent. As a partner driver, now a worker, I do not know anymore when and if Uber is charging a "surge" price or not, if no driver is picking up the trip request within a reasonable timeframe. Drivers are completely being kept in the dark as workers, without safe and effective protection measures that would meet their purpose. I can no longer foresee my earnings nor plan ahead for the future anymore. The only thing I can control is the contract between Uber and the rider. If before being a private hire driver on the Uber platform was a healthy reliable opportunity, now it's become an enigma, a survival of the most resilient to stress, as Uber does not allow me to earn more than £700-£900 per seven days of driving. I cannot afford to have time off anymore, whereas before weekends were optional, if and when I needed more money for investments or for my other ventures. Now I have to plan 1-2 days off a month very carefully, while practically having my hands forced by Uber's new business model to decline between 2,000-3,000 trips every month due to them being underpaid. This is another aspect of concern regarding this business model: most drivers are physically and mentally exhausted (not having sufficient resources to plan their lives at ease). I am **[Your Age]** years old and I cannot plan anything in my life anymore, compared to 2019 when I was earning fair and decent money, which I was using to plan for my future. We have all been through the pandemic and its devastating effects, and private hire drivers did not deserve such mistreatment whatsoever.

18. It is now a huge opportunity for Uber, for the corporation to make more money than ever before and satisfy its investors' expectations. Hiding behind the fact that they are keeping more from the fares paid by riders to fulfil the worker benefits, it reflects into taking advantage of drivers like never before. I personally opted out from paying into a pension pot because I do not believe in pension, and that is my right. I would have to live and work an extra 250 years to be able to accumulate enough

money in the pot to secure my minimal survival past 67 years, the pension age. Also, I can never trust anybody else with my money, especially pension funds that will most likely crash and lose all of it. It happened before and the likelihood of happening again is just a matter of time. At present, I cannot opt out of the weekly holiday pay, but that means nothing compared to what I used to make to pay for my holidays. Nowadays, I cannot afford to book days off to go on holiday because I would run out of money in a few days. In 2019, I could easily accumulate money to cover unforeseen events. Despite informing Uber that I do not accept these benefits as I did not agree to these changes, they are still withholding money from the fares of my riders to pay other drivers' benefits and the pension fund manager, Adecco. This is totally unfair for me personally, as I should have a say in this regard. The partner drivers who did not like the previous business model should have left the platform. The main parties of the Respondents actually did it, even before destroying everything for all other drivers. There is no Uber representative who I could talk to about this state of facts. Uber executives are only talking and negotiating with these unions that have profoundly affected my business, unions that I did not delegate to represent my interests for that matter. All implemented changes should only apply to their members at most. Plus, since I do not agree to delegate my legal representation, Uber should invite me to all these closed doors negotiations. This does not happen and Uber would never agree to me being present. This is a complete cancellation and disregard to my legal rights as an individual.

19. If the ruling was to be applied to the Respondents only, I would not have been in this state of affairs, the business model probably would have stayed the same, as the main Respondents are not private hire drivers in partnership with Uber, nor have they been throughout the legal proceedings. How they managed to pay for their daily living without working, when they affirmed the pay was not sufficient when they were actively driving with Uber, is still an enigma for me. This equates to and ticks all the points of an external influence to purposely destabilise a whole industry and harm millions that depend on the Uber platform throughout the UK.

20. If the business model does not go back to treating drivers as self-employed, they will undoubtedly become modern slaves working for rental companies and ride hailing corporations. This will also slow down the switch to electric, as no drivers partnering with Uber today can afford an electric vehicle, as even renting one is an unfeasible hurdle. The drivers who switched to electric vehicles are living on the brink of survival, highly likely struggling to make ends meet given the cost of living crisis. Despite the number of drivers using

electric vehicles and making the switch, about 25% to 50% of them are taking the vehicles back, having no support from these corporations or the government. As opposed to this, the black cab drivers received about £50,000 in non-refundable grants from the government to change their vehicles overnight, considering their earnings ranging between £3,000-£5,000 per week. When comparing the two types of drivers, it is obvious that one side is constantly being dragged to the ground and disadvantaged, whereas the other part is thriving as it is supported and incentivised by everyone. I am not against any type of taxi or person making a lot of effort to earn a living, but I did not become a private hire driver to **steal** their money. There is enough business to go around for everybody. Not having any support from the government despite providing a high quality and comfortable service, cumulated with and consolidated by the Supreme Court's ruling, have caused me tremendous harm as well as affected my overall wellbeing. To exhaustively comprehend the matter at hand, as a result of continuously working throughout 2020, during the covid-19 pandemic, I have not received any hazard payment for putting my health at high risk, no bonuses, not even a thank you from the government for directly/indirectly saving lives.

21. There is no clear rule book to define as to how much private hire operators are allowed to

withhold from the total fare paid by riders, their justified need to withhold a certain amount, or as to why drivers that decline the benefits offered to them by Uber should also pay for other drivers' benefits. Ride hailing platforms would argue here that those partner drivers who do not agree to these changes, should not be permitted to work on the platform, but isn't that what should have happened with the Respondents? Shouldn't they have given up working with Uber or others if they didn't like the system, instead of ruining the business of the majority self-employed drivers by bringing such an outrageously bogus case to Court? To me, the suspicious and outstanding situation here is that all three main Respondents, Mr Aslam, Mr Farrar, and Mr Dawson were not active partner drivers on the Uber platform. I strongly believe, based on factual arguments, that they were not driving on any platform at the time, and they didn't return to partnering with any licensed private hire operator after the case was ruled in their favour. They stopped before filing the lawsuit. Another suspicious fact is that they have dissolved their union trading as ADCU (App Drivers & Couriers Union), but they are still acting as active and fighting for drivers, having a media presence even in the current situation.

22. At present, the new business model with partner drivers being considered workers and paid less and less per mile on most trips, is detrimental to their

health, wellbeing and the overall economy. The private hire sector is now large enough to have a significant impact on the overall economy. Before the Court's ruling, over 75% of the money (including tips) earned on the Uber platform was going into the pockets of active drivers. Most of that over 75% was going back into the active economy, and a small portion of the rest towards National Insurance and income tax (drivers, Uber employees, corporation tax, etc.).

23. Nowadays, Uber can set the prices higher than 2019, but as opposed to this, I am most times paid substantially less per mile than before. Much more common now, riders are complaining about the high fares, while drivers are getting paid too little. As an example, Uber offered me one trip for which the rider was paying £149 and I was getting paid only £92. I found out, after breaking down the numbers, about trips when Uber paid me only 48%-49% from the total fare paid by riders. It is like a Wild West in the market, with the power in the hands of these corporations and TfL. Drivers have no say anymore, and the transparency was reduced to zero. There must have been a collusion against private hire drivers, between TfL, the taxi industry, the unions and Uber, because if Uber presented this business model when it first applied for a licence in 2012, TfL most likely would've refused them a licence to operate in London as such. Nobody is willing to talk about the effects and the consequences of the Court's ruling on my earnings, nor they expressed any interest in listening to me, and that is concerning to see that the regulator, who is supposed to support my wellbeing, is silent in these matters and totally unapproachable.

24. As a consequence, at present time, taxi operators are charging riders between 2 to 5 times more than what Uber does. I have had numerous riders complaining about the taxi prices, when no Uber driver was accepting their booking for long periods of time. Taxi operators know what is happening in the market and are taking advantage of every opportunity, which is affecting even disabled people and critical workers, like doctors and nurses, in dire need to go from A to B. The ruling to treat me as worker has had deep implications in the economy, and it destabilised millions needing affordable transportation, as Uber and self-employed private hire drivers have brought private rides and reliability to the people. Prior to Uber, the ride sharing industry used to be a luxury service available only to the rich or the ones with more money than average people. With private hire drivers treated as workers, the industry is being shifted back to an even bigger gap between classes of people, with a higher discrepancy than before the service started to be available to everyone.

25. The private hire industry, and Uber predominantly, is so important that I had British MPs as riders, as well as US senators on their way to Westminster. I strongly believe that I deserve to be allowed to be independent and earn as much money as I can in the least amount of time, if and when available in the market, as well as to have the power to hold the ride sharing platform of my choice accountable for real issues that were never brought to Court or to the attention of TfL in all these years. All problems with Uber, outside the object of this case, have been fabricated and they were actually failings of TfL, the regulator. There were vast misinformation campaigns, pushed forward by the media and TfL, about Uber drivers and the private hire industry in general. I am a good citizen and have proven myself in the market over the years as a highly skilled driver.

26. Nowadays, in many instances, Uber is paying me less than what a person would pay to travel by bus. This should never be the case, as I am providing a private hire service with exceptional comfort, safety, and peace of mind that exceeds and rules out the quite often inconvenience of a journey by bus, with riders driven precisely from door to door. Be it as it may, the carbon footprint of one such trip is much lower compared to a bus, as I am driving an almost negligible emissions vehicle.

27. In 2019, Uber offered huge bonuses, worth of up to £7,000, to qualifying partner drivers who completed over 10,000 trips on the platform, times when drivers were paying only 25% service fee. Back then, Uber had money to spare, but nowadays it appears that it needs more and more from drivers. This is the aftermath and the result of drivers being treated as workers by these corporations, when I cannot be and should never be a worker for all the reasons mentioned in this letter.

28. Before changing its business model, Uber was decent, reliable and good when I was a contractor, and that helped me and my family in many ways. It is worth mentioning that I used to make enough money to take care of and improve my health, could afford to do the same for my family, I was able to purchase a high specification vehicle in order to improve comfort, could afford to live in a better house and in a better area, could book days and even weeks off in order to travel, could make plans to buy a house of my own, I could make investments (my pension) and be in total control of my life, and most of all, I could start my own company. Today, my projects for the future are in danger due to my overall income dropping a whopping 50% or more, and having to work seven days a week. Given the present circumstances, I have no more money to expand my business further, and that will translate into a less diverse economy if my projects becomes unfeasible. In short, I was thriving as a self-employed private hire driver and could plan for the future, the way I wanted. The fact that I am now treated as worker, paved the way for these corporations to pocket exponentially more money, to literally rip off the riders and drivers that have kept their business alive during the worst pandemic of the last 50 years.

29. In the light of what was brought to knowledge, the same Employment Appeals Tribunal decided that a taxi driver was not a worker when picking up riders on the "Mytaxi" app, which is an identical service in the market as Uber. There is absolutely no difference. The fact that I cannot call the rider after the trip has ended was a prerequisite to getting licensed by TfL, but an unnecessary one, as all drivers are being vetted and validated by TfL, and not by Uber, same as taxi drivers. Historically, there have been more taxi drivers acting unlawfully than private hire drivers, due to the more secure business model of the latter. All in all, what taxi (nowadays working with ride hailing platforms) drivers and private hire drivers are offering is the same, transportation from A to B via an intermediary platform. Both groups cannot be workers, no matter what options are available on the platform.

30. As of 28 November 2022, Uber started altering my online time on the platform, by only showing me the driving time, excluding the total time spent

online on the app to accept offers, most likely knowing that these unlawful unions are trying to achieve a minimum payment from the moment drivers go online, in order to compensate their unaccounted time of online availability, without which Uber per se could not operate. The very same unions that created these unwanted issues, are now trying to create more destruction, by causing more harm to drivers. What they are asking for is absurd, considering the global implications, therefore the unions' actions should be treated as illegitimate intentions. Now for instance, in contradiction to being online for about eight hours and 40 minutes, in the daily summary on the Uber app, I'm being shown only five hours and 27 minutes, to make it look as if I'm earning £30-£40 per hour constantly. Uber went as far as altering the time of online availability on the platform for all the weeks from the past. This is an example of how these companies are trying to mould themselves on a future situation, pressured by the union actions I do not support nor consented to. Overall, I truly believe that unions for the self-employed should not exist in the first place. I know and am aware that it is a legal right to form a union, but what about the ones that prefer taking legal action independently, if necessary and wish to do so? In my view, unions are open to corruption, and I am absolutely positive that this is the case with what has happened so far, as the unions involved in these matters have a minority of drivers as members. Being self-employed or employee is a personal choice and should remain as such, as both come with risks and benefits. Whoever does not agree with the status should change careers, and not ruin the business of others. In this case, the Respondents did leave the platform, but left a heritage that crippled all others in the industry. That is not fair in any way and should be treated by the Court as malicious intent as well as contempt of Court. Right from the beginning, their requests and arguments have not matched the reality whatsoever, and I can prove it with my own activity on the Uber platform.

31. I am the only one interacting directly with the person making the booking on the Uber platform, and have absolute power to agree with all changes to the initial booking, or end it midway and at any time, without agreeing to the changes requested by the rider, when and if circumstances such as rider's behaviour and actions towards me or my vehicle, or driving to a dangerous area require to do so. Uber has no say nor power over what I can accept or not, even during the contract, as it does not physically interact with its customers, therefore the present contract between Uber and the rider (following a previous Court's ruling) is in reality a contract between the rider and the driver alone due to its aforementioned aspects. Consequently, the rider can only contract with me, as I am spending my time driving from A to B with my own means,

paying for my vehicle's running costs, as well as having a percentage taken by Uber from the total fare paid by the rider. This is why I need to know and be shown in the app, exactly how much the rider is being charged by the platform for each change added during the contract or before the start of the contract, so I can decide if the offer is worth confirmation and completion. What Uber is doing now, in the light of the ruling, highly likely resembles fraud and could lead to lawsuits for audit, compensation, as well as change of its business model. It goes without saying that any lawsuit brought against Uber means another big gap in my personal finances. I refuse to link with others during a lawsuit, because I have been betrayed by partners (unions) in the past when have lost control of my case and cause.

32. During the legal proceedings I did not have legal representation, as I have never delegated a union to represent my interests, nor have I been contacted about these matters, and because I have been working on the platform as one of the best drivers in London, it was imperative to be consulted regarding any possible change to my business. I have earned through my outstanding service the right to be a stakeholder in Uber, therefore to be legally asked for an opinion with regards to these matters. My person is a legal entity, and as such, what has been brought in front of the Court is a breach of my basic rights as a self-employed individual. Nobody had and has the right to disrupt my business without my consent. At most, the Plaintiff could've asked that the changes apply exclusively to them. The fact that they didn't, and left the platform, is at least questionable and disrespectful toward the Court and everyone else. I don't even know if they have asked this in Court or not. If they did not, at any stage of the proceedings, the breach of my rights is clear. Also, I do not know how many the "others" forming the Plaintiff were there. From what I know for a fact, most private hire drivers are not members of any union. There are countries as Australia, where high courts of law have decided that the changes apply only to the Plaintiff. In my opinion such a ruling would've been more than fair and legally comprehensive.

33. Upon studying the worker law in detail, I have understood that the ruling of the Supreme Court was given based on an incomplete interpretation, whereas a law must be interpreted exhaustively with no exceptions, as all aspects and facts have to to be satisfied in order to issue a comprehensive and impartial judgement. My findings led me to believe that as a private hire driver or taxi driver partnering with ride sharing platforms, I can never be a worker, based on the followings:

    A: **a worker must show up and do the work;**
    Based solely on this point, I cannot be forced to show up and do the work for obvious safety

reasons. There are many dangerous riders out there and as an example, I had passengers that wanted to fight me or who touched the controls of my vehicle while I was driving, putting everyone in danger. In other instances, I had to terminate the trip and ask the riders to step out of the vehicle, or have decided to not start it when seeing the state of the person. The reasons as to why I cannot be forced to work are numerous, therefore I am not a worker. Conversely, a worker cannot decide where and when to do the work.

B: **the employer/the work provider, Uber, has to provide the tools for me to do the work;**

This can never happen, as Uber will never pay for all my costs to work on the platform.

C: **there is no contract of employment between Uber and me, but only a Terms & Conditions Agreement that can be terminated unilaterally at any point in time by either party, without any liability.**

Based on the above arguments, I am respectfully asking the Employment Appeals Tribunal of the UK Supreme Court to rescind the decision to treat private hire drivers in partnership with Uber as workers, to give me back the freedom to earn a decent living and to be able to ask for compensation from my business partners for the damages caused to me and my family as a cause of its implementation. I can never be a worker as stated by the full extent of the workers law. As I have never been a worker when driving with Uber, this case should've never made the object of an Employment Tribunal. I have not agreed with these changes, and their repercussions have damaged my life irreparably, while the individuals forming the Plaintiff are bragging in the media about how "cool" they are for taking Uber down. My request is also supported and reinforced by the Plaintiff never returning to partner with Uber after winning the case.

[Your Signature]

[Your Name]

The Big Conclusion

So, who is to blame for this mess?

Number 1: the ADCU founders and "others" that brought the case to court.
These are the main culprits that showed total disrespect toward every private hire/independent contractor that partnered with Uber and its competition. They are all crooks, and if you look into their personal histories, we guarantee that you'll find shocking information. There is a high chance of at least one of them being a member of a secret society, an undercover government agent, or both. In other words, a "legal" disruptor engaged in disrupting the transportation industry to favour groups of interest and the new world order (poorer people), the green agenda, Agenda 21, etc. The founders of ADCU have been paid to set everybody up. Otherwise, they would've asked that these changes apply only to them, and to the ones that requested it later on.

Number 2: the UK's Courts of Law.
The UK's High Court and The Supreme Court have misinterpreted the workers law and issued the wrong decisions. This is unacceptable for the highest courts in a country such as the UK. To decide on the case, the justices have only interpreted a part of the workers law, when it should be interpreted in its entirety. The text of the law must be exhausted in order to issue a comprehensive and fair judgement. On top of that, the courts knew that a very small

minority of drivers have brought the case to court. The initiators were not drivers! This should've raised concerns straight away. In addition, since the case has involved all drivers by default (you were always part of the case by default because the contentious was affecting you directly), the judges should've subpoenaed all other drivers (the majority). Even before this, people that did not drive on any platform at the time of filing the lawsuit, should have not been allowed to proceed with such a case. These are some of the main aspects that made us to believe that there was a higher influence over the courts of law, to issue judgements against existing laws meant to support overregulation. At most, the judgement should have applied only to the Plaintiff (Yaseen Aslam, James Farrar, Robert Dawson, and others). The fact that the High Court and The Supreme Court did not mention this in their judgements is very peculiar and unfair.

Number 3: Uber.

Uber had two absolute choices: to enforce the decision only on the Plaintiff, or not enforce it at all. The Supreme Court decision has the power of law but it is not above the existing laws passed by Parliament (the Parliament has power over the Supreme Court) and individual choice. The fair stance Uber should have adopted, was to ask all drivers about what they want and listen to them individually (they –Uber– asked but they didn't listen to drivers' replies). What they have figured out on the way was an opportunity to grab more money at the expense of its drivers and riders, under the umbrella of the Supreme Court decision. The previous system is the only one that can work with Uber: 75% or more paid to drivers and 25% or less to Uber. This is how Uber grew into a tech giant. There is no other way because that was a proven system. With the current business model, Uber would've never become what it is today. Uber could've just applied the Supreme Court's decision after consulting and agreeing with all its partners, or continue as normal. It gave in to regulator and crooked unions pressures instead, joining the overall collusion by naming executives with questionable track records and affiliations (Andrew Brem was an executive at British Airways when the company stopped giving free drinks/snacks during the flight and work conditions became worse). We had the chance to speak with union leaders from British Airways, who confirmed Andrew Brem's total disregard toward staff and business partners wellbeing when at the airline.

Number 4: Transport for London and The Mayor's Office.

It is outrageous and highly unbecoming that these two institutions have openly lobbied against private hire from the start. This does not make it a free economy! These two, which are actually acting as one, could be pointed out as the main culprits for what's happening today with Uber. Let's say they don't care about private hire drivers (which is totally not democratic), but they don't care about riders either. With public transport severely disrupted and unreliable in many parts of London (even central), and extortionate taxi fares (taking advantage of what is happening with Uber), TfL and Sadiq Khan are probably the biggest implementers of the new world agenda in the UK. The problem we're all experiencing is that there is no higher authority to complain to that would listen. We have the Parliament, but since it brims with conspirators and rich actors promoting depopulation, how could we find justice in a snake pit? What's left to do, is to take legal matters into our own hands and either crash/change the entire system on our own terms, or suffer and perish on their own terms. Most people in London know that TfL is an outdated and prone to corruption system, and see the current Mayor as part of the problem, but no one does anything. They're being allowed to destroy life through much more absurd regulations and charges (money for nothing and out of thin air), as they turned London into a sanctions hell hole. How can a regulator licence two drivers, but charge only one of them for driving into central London and airports, although they're driving the same type of vehicle and offer the same type of service (most times, private hire is way better)? This needs to stop, as it is against a freer and fairer future, and against our plans to protect nature.

Number 5: drivers.

There were a few drivers that joined the legal action, despite their lack of education in law. We are pretty sure that some of them were paid to act like so. For not consulting all other drivers, and colluding behind their backs, we have no gratitude for them. Nevertheless, this book is the beginning of something very big, the biggest positive change in private hire since its inception. We will succeed in bringing about the right world order (based on personal freedom, private property, and privacy), and hold the ones to blame for our destruction responsible in courts of law. However this takes time, education, precision, and sophistication.

As you've reached the end of this book, never forget that cumulative legal action against the grand destructive narrative (without higher support) is weak and vulnerable to corruption, while individual collective action is the armageddon for our delegates (authorities) and the corruptible justice system. Start your own if you feel courageous enough, and let's meet in the middle, driven by the same purpose. Then we shall go back to our homes victorious, and wink at each other when crossing paths, in sign that we are strong and smart individuals united in purpose, not in a union...

The Legal Way And Who To Contact

You need to contact the following institutions and entities: Uber (ask them to go back to the old system, head offices – ask to sit down with the executives for discussions without any unions), The Supreme Court, Transport for London, The Mayor of London (just imagine Sadiq's reaction when receiving tens of thousands of requests to relicense Uber with the old system (the contract to be set between the rider and the driver), your MP for support in these matters (you are the local business), and lastly, the Parliament (send your request and start a petition on the UK Government website (or sign our one). The Parliament is the highest institution in the UK that can change the law and override Supreme Court judgements. You need to prove all these institutions, that the majority does not agree with the Supreme Court decision. A very small minority (probably less than 100 drivers) brought the case to court and severely damaged the business for tens of thousands of self-employed individuals.

Stand up for yourself or lose your rights forever!

Last Hour Developments

While finishing this book we have received an answer from The Supreme Court stating the following:

"We are unable to offer comments on decisions reached by the Supreme Court Justices when operating in a judicial capacity.

I should also explain that there is no provision for reviewing Supreme Court judgements in the way you have requested.

I am sorry I am unable to help you any further on this occasion."

Registry Manager

The above reply to our arguments and request regarding the Supreme Court decision on the case, means that no Judge of The Supreme Court saw the papers. In addition, we did find provisions regarding decisions of The Supreme Court, which stated that if a decision of the Justices has caused you harm, you should contact the Court. We still believe that all drivers should send the same request to the court, in order to put things into motion. We received this response as we are not professional lawyers, and the judicial system is seeing people that are representing themselves as inexperienced, unworthy individuals to consider or listen to. The response also contains a link to the case, including the video of the two sessions (21-22 July 2021). In that video, we had our hunch confirmed: *"Aslam and others"* presented the situation, as if all drivers on the Uber platform wanted to be workers, which was totally incorrect and unfair. This was the Appeal stage, but it is correct to think that it was the same during the High Court proceedings. Also, our request should've raised concerns as we had been part of the proceedings by default, as these legal matters concerned as well as affected us directly. We'll proceed to sending our documents (amended/shorter text) to each individual Justice that was part of the decision.

The Supreme Court reply enforces our belief that other, more powerful and influential forces were behind the judgement. We did not want to be treated as workers, but someone decided for us, without asking for permission, and their ruling changed our lives on the platform in an almost irreparable way. This is why we think that all drivers should take *"Aslam and others"* to court for financial and statute damages. Nobody, and absolutely nobody, is allowed to take control of self-employed individuals' business without their individual permission. The Supreme Court Justices know this, and their decision was given against existing laws, or at least an incorrect or incomplete misinterpretation of the workers law. Send your papers to the court and ask for the same changes.

The moment was ripe for a book like this to reach the masses, as just complaining without action is cowardice and ignorant. Always take the legal way! We have studied law for many years, and what we found out is that there is always a way to everything; you just need to find it through knowledge.

A Campaign To Raise Awareness For PEACE

There will be no world war III
NFT

Total Supply: 30 Billion

Price: £10

We predict at least 30 billion naturally healthy and free people living in PEACE on planet Earth by 2050.

Money raised will be used to build self-sustained homes integrated into nature, and a new economy without money of any kind.

You can purchase more than one NFT, but please keep only one for yourself and share the others with your family and friends. We want all good men & women to own one in order for our campaign to work and be successful.

Purchase this unique NFT via **www.arkdian.com**.

References

https://www.dailymail.co.uk/news/article-10466009/amp/Electric-cars-green-hoped-polluting-particles-warns-Environment-secretary.html

https://www.supremecourt.uk/cases/uksc-2019-0029.html

Printed in Great Britain
by Amazon